RESILIENCE ENGINEERING

Resilience Engineering
Concepts and Precepts

Edited by
ERIK HOLLNAGEL
Linköping University, Sweden
DAVID D. WOODS
Ohio State University, USA
NANCY LEVESON
Massachusetts Institute of Technology, USA

CRC Press
Taylor & Francis Group
Boca Raton London New York

CRC Press is an imprint of the
Taylor & Francis Group, an **informa** business

CRC Press
Taylor & Francis Group
6000 Broken Sound Parkway NW, Suite 300
Boca Raton, FL 33487-2742

© 2006 by Erik Hollnagel, David D. Woods and Nancy Leveson.
CRC Press is an imprint of Taylor & Francis Group, an Informa business

No claim to original U.S. Government works

Printed on acid-free paper
Version Date: 20160226

International Standard Book Number-13: 978-0-7546-4641-9 (Hardback) 978-0-7546-4904-5 (Paperback)

Visit the Taylor & Francis Web site at
http://www.taylorandfrancis.com

and the CRC Press Web site at
http://www.crcpress.com

Contents

Preface

Decades of efforts aimed at understanding what safety is and why accidents happen have led to several significant insights. One is that untoward events more often are due to an unfortunate combination of a number of conditions, than to the failure of a single function or component. Another is that failures are the flip side of successes, meaning that there is no need to evoke special failure mechanisms to explain the former. Instead, they both have their origin in performance variability on the individual and systemic levels, the difference being how well the system was controlled.

It follows that successes, rather than being the result of careful planning, also owe their occurrence to a combination of a number of conditions. While we like to think of successes as the result of skills and competence rather than of luck, this view is just as partial as the view of failures as due to incompetence or error. Even successes are not always planned to happen exactly as they do, although they of course usually are desired – just as the untoward events are dreaded.

A case in point is the symposium behind this book. As several of the chapters make clear, the notion of resilience had gradually emerged as the logical way to overcome the limitations of existing approaches to risk assessment and system safety. Ideas about resilience had been circulated more or less formally among several of the participants and the need of a more concerted effort was becoming obvious. Concurrently, a number of individuals and groups in the international community had begun to focus on a similar class of problems, sometimes talking directly about resilience and sometimes using related terms. In the USA, the Santa Fe Institute had begun programmes on robustness in natural and engineering systems and on robustness in social processes. Within the school of high-reliability organizations, the term *resilience* appeared in paper titles, e.g., Sutcliffe & Vogus (2003). A related concept was the proposal of a conceptual framework, named Highly Optimised Tolerance (HOT), to study fundamental aspects of complexity, including robust behaviour (e.g., Carlson & Doyle, 2000 & 2002). In Europe, a research organization of scientists and practitioners from many disciplines collaborated to explore the dynamics of social-ecological systems under the name of the Resilience Alliance, while another group was called the Information Systems for Crisis Response and Management or ISCRAM community.

The intention of getting a group of international experts together for an extended period of time to discuss resilience was, however, just one component. Some of the others were that one of the protagonists (David D. Woods) was going to spend some time in Europe, that initial inquiries indicated that the basic funding would be available, and that most of the people whom we had in mind were able and willing to interrupt their otherwise busy schedules to attend the symposium.

The symposium itself was organized as a loosely structured set of discussions with a common theme, best characterized as long discussions interrupted by short presentations – prepared as well as *ad hoc*. The objective of the symposium was to provide an opportunity for experts to meet and debate the present and future of Resilience Engineering as well as to provide a tentative definition of organizational resilience. Readers are invited to judge for themselves whether these goals were achieved and whether the result is, indeed, a success. If so, the credit goes to the participants both for their willingness to take part in a process of creative chaos during one October week in Söderköping, and for their discipline in producing the written contributions afterwards.

We would also like to thank the two main sponsors, the Swedish Nuclear Power Inspectorate (SKI) and the Swedish Civil Aviation Administration (LFV), who were willing to support something that is not yet an established discipline. Thanks are also due to Kyla Steele and Rogier Woltjer for practical and invaluable assistance both during the symposium and the editing process.

Linköping, July 2005

Erik Hollnagel   David D. Woods   Nancy G. Leveson

Prologue: Resilience Engineering Concepts

David D. Woods
Erik Hollnagel

Hindsight and Safety

Efforts to improve the safety of systems have often – some might say always – been dominated by hindsight. This is so both in research and in practice, perhaps more surprising in the former than in the latter. The practical concern for safety is usually driven by events that have happened, either in one's own company or in the industry as such. There is a natural motivation to prevent such events from happening again, in concrete cases because they may incur severe losses – of equipment and/or of life – and in general cases because they may lead to new demands for safety from regulatory bodies, such as national and international administrations and agencies. New demands are invariably seen as translating into increased costs for companies and are for that reason alone undesirable. (This is, however, not an inevitable consequence, especially if the company takes a longer time perspective. Indeed, for some businesses it makes sense to invest proactively in safety, although cases of that are uncommon. The reason for this is that sacrificing decisions usually are considered over a short time horizon, in terms of months rather than years or in terms of years rather than decades.)

In the case of research, i.e., activities that take place at academic institutions rather than in industries and are driven by intellectual rather than economic motives, the effects of hindsight ought to be less marked. Research should by its very nature be looking to problems that go beyond the immediate practical needs, and hence address issues that are of a more principal nature. Yet even research – or perhaps one should say: researchers – are prone to the effects of hindsight, as pointed out by Fischhoff (1975). It is practically a characteristic of human nature – and an inescapable one at that – to try to make sense of what has happened, to try to make the perceived world comprehensible. We are consequently constrained to look at the future in the light of the past. In this way our experience or understanding of what has happened inevitably colours our anticipation and preparation for what could go wrong and thereby holds back the requisite imagination that is so essential for safety (Adamski & Westrum, 2003). Approaches to safety and risk prediction furthermore develop in an incremental manner, i.e., the tried and trusted approaches are only changed when they fail and then usually by adding one more factor or element to account for the unexplained variability. Examples are easy to find such as 'human error', 'organisational failures', 'safety culture', 'complacency', etc. The general principle seems to be that we add or change just enough to be able to explain that which defies the established framework of explanations. In contrast, resilience engineering tries to take a major step forward, not by adding one more concept to the existing vocabulary, but by proposing a completely new vocabulary, and therefore also a completely new way of thinking about safety. With the risk of appearing overly pretentious, it may be compared to a paradigm shift in the Kuhnian sense (Kuhn, 1970).

When research escapes from hindsight and from trying merely to explain what *has* happened, studies reveal the sources of resilience that usually allow people to produce success when failure threatens. Methods to understand the basis for technical work shows how workers are struggling to anticipate paths that may lead to failure, actively creating and sustaining failure-sensitive strategies, and working to maintain margins in the face of pressures to do more and to do it faster (Woods & Cook, 2002). In other words, doing things safely always has been and always will be part of operational practices – on the individual as well as the organisational level. It is, indeed, almost a biological law that organisms or systems (including organisations) that spend all efforts at the task at hand and thereby neglect to look out for the unexpected, run a high risk of being obliterated, of meeting a speedy and unpleasant demise. (To realise that, you only need to look at how wild birds strike a balance between head-down and head-up time when eating.) People in their different roles within an organisation are aware of potential paths to failure and therefore develop failure-sensitive strategies to forestall these possibilities. Failing to do that brings them into a reactive mode, a condition of constant fire-fighting. But fires, whether real or metaphorical, can only be effectively quelled if the fire-fighters are proactive and able to make the necessary sacrifices (McLennan et al., 2005).

Against this background, failures occur when multiple contributors – each necessary but only jointly sufficient – combine. Work processes or people do not choose failure, but the likelihood of failures grow when production pressures do not allow sufficient time – and effort – to develop and maintain the precautions that normally keep failure at bay. Prime among these precautions is to check all necessary conditions and to take nothing important for granted. Being thorough as well as efficient is the hallmark of success. Being efficient without being thorough may gradually or abruptly create conditions where even small variations can have serious consequences. Being thorough without being efficient rarely lasts long, as organisations are pressured to meet new demands on resources. To understand how failure sometimes happens one must first understand how success is obtained – how people learn and adapt to create safety in a world fraught with gaps, hazards, trade-offs, and multiple goals (Cook et al., 2000).

The thesis that leaps out from these results is that failure, as individual failure or performance failure on the system level, represents the temporary inability to cope effectively with complexity. Success belongs to organisations, groups and individuals who are resilient in the sense that they recognise, adapt to and absorb variations, changes, disturbances, disruptions, and surprises – especially disruptions that fall outside of the set of disturbances the system is designed to handle (Rasmussen, 1990; Rochlin, 1999; Weick et al., 1999; Sutcliffe & Vogus, 2003).

From Reactive to Proactive Safety

This book marks the maturation of a new approach to safety management. In a world of finite resources, of irreducible uncertainty, and of multiple conflicting goals, safety is created through proactive resilient processes rather than through reactive barriers and defences. The chapters in this book explore different facets of resilience as the ability of systems to anticipate and adapt to the potential for surprise and failure.

Until recently, the dominant safety paradigm was based on searching for ways in which limited or erratic human performance could degrade an otherwise well designed and 'safe system'. Techniques from many areas such as reliability engineering and management theory were used to develop 'demonstrably safe' systems. The assumption seemed to be that safety, once established, could be maintained by requiring that human performance stayed within prescribed boundaries or norms. Since 'safe' systems needed to include mechanisms that guarded against people as unreliable components, understanding how human performance could stray outside these boundaries became important.

According to this paradigm, 'error' was something that could be categorised and counted. This led to numerous proposals for taxonomies, estimation procedures, and ways to provide the much needed data for error tabulation and extrapolation. Studies of human limits became important to guide the creation of remedial or prosthetic systems that would make up for the deficiencies of people. Since humans, as unreliable and limited system components, were assumed to degrade what would otherwise be flawless system performance, this paradigm often prescribed automation as a means to safeguard the system from the people in it. In other words, in the 'error counting' paradigm, work on safety comprised protecting the system from unreliable, erratic, and limited human components (or, more clearly, protecting the people at the blunt end – in their roles as managers, regulators and consumers of systems – from unreliable 'other' people at the sharp end – who operate and maintain those systems).

When researchers in the early 1980s began to re-examine human error and collect data on how complex systems had failed, it soon became apparent that people actually provided a positive contribution to safety through their ability to adapt to changes, gaps in system design, and unplanned for situations. Hollnagel (1983), for instance, argued for the need of a theory of action, including an account of performance variability, rather than a theory of 'error', while Rasmussen (1983) noted that 'the operator's role is to make up for holes in designers' work.' Many studies of how complex systems succeeded and sometimes failed found that the formal descriptions of work embodied in policies, regulations, procedures, and automation were incomplete as models of expertise and success. Analyses of the gap between formal work prescriptions and actual work practices revealed how people in their various roles throughout systems always struggled to anticipate paths toward failure, to create and sustain failure-sensitive strategies, and to maintain margins in the face of pressures to increase efficiency (e.g., Cook et al, 2000). Overall, analysis of such 'second stories' taught us that failures represented breakdowns in adaptations directed at coping with complexity while success was usually obtained as people learned and adapted to create safety in a world fraught with hazards, trade-offs, and multiple goals (Rasmussen, 1997). In summary, these studies revealed:

- How workers and organisations continually revise their approach to work in an effort to remain sensitive to the possibility for failure.
- How distant observers of work, and the workers themselves, are only partially aware of the current potential for failure.
- How 'improvements' and changes create new paths to failure and new demands on workers, despite or because of new capabilities.
- How the strategies for coping with these potential paths can be either strong and resilient or weak and mistaken.
- How missing the side effects of change is the most common form of failure for organisations and individuals.
- How a culture of safety depends on remaining dynamically engaged in new assessments and avoiding stale, narrow, or static representations of the changing paths (revising or reframing the understanding of paths toward failure over time).
- How overconfident people can be that they have already anticipated the types and mechanisms of failure, and that the strategies they have devised are effective and will remain so.
- How continual effort after success in a world of changing pressures and hazards is fundamental to creating safety.

In the final analysis, safety is not a commodity that can be tabulated. It is rather a chronic value 'under our feet' that infuses all aspects of practice. Safety is, in the words of Karl Weick, a dynamic non-event. Progress on safety therefore ultimately depends on providing workers and managers with information about changing vulnerabilities and the ability to develop new means for meeting these.

Resilience

Resilience engineering is a paradigm for safety management that focuses on how to help people cope with complexity under pressure to achieve success. It strongly contrasts with what is typical today – a paradigm of tabulating error as if it were a thing, followed by interventions to reduce this count. A resilient organisation treats safety as a core value, not a commodity that can be counted. Indeed, safety shows itself only by the events that do not happen! Rather than view past success as a reason to ramp down investments, such organisations continue to invest in anticipating the changing potential for failure because they appreciate that their knowledge of the gaps is imperfect and that their environment constantly changes. One measure of resilience is therefore the ability to create foresight – to anticipate the changing shape of risk, before failure and harm occurs (Woods, 2005a).

The initial steps in developing a practice of Resilience Engineering have focused on methods and tools:

• to analyse, measure and monitor the resilience of organisations in their operating environment.
• to improve an organisation's resilience vis-à-vis the environment.
• to model and predict the short- and long-term effects of change and line management decisions on resilience and therefore on risk.

This book charts the efforts being made by researchers, practitioners and safety managers to enhance resilience by looking for ways to understand the changing vulnerabilities and pathways to failure. These efforts begin with studies of how people cope with complexity – usually successfully. Analyses of successes, incidents, and breakdowns reveal the normal sources of resilience that allow systems to produce success when failure threatens. These events and other measures indicate the level and kinds of brittleness/resilience the system in question exhibits. Such indicators will allow organisations to develop the mechanisms to create foresight, to recognise, anticipate, and defend against paths to failure that arise as organisations and technology change.

Part I: Emergence

Chapter 1

Resilience – the Challenge of the Unstable

Erik Hollnagel

Safety is the sum of the accidents that do not occur. While accident research has focused on accidents that occurred and tried to understand why, safety research should focus on the accidents that did not occur and try to understand why.

Understanding Accidents

Research into system safety is faced with the conundrum that while there have been significant developments in the understanding of how accidents occur, there has been no comparable developments in the understanding of how we can adequately assess and reduce risks. A system is safe if it is impervious and resilient to perturbations and the identification and assessment of possible risks is therefore an essential prerequisite for system safety. Since accidents and risk assessment furthermore are two sides of the same coin, so to speak, and since both are constrained in equal measure by the underlying models and theories, it would be reasonable to assume that developments in system safety had matched developments in accident analysis. Just as we need to have an aetiology of accidents, a study of possible causes or origins of accidents, we also need to have an aetiology of safety – more specifically of what safety is and of how it may be endangered. This is essential for work on system safety in general and for resilience engineering in particular. Yet for reasons that are not entirely clear, such a development has been lacking.

The value or, indeed, the necessity of having an accident model has been recognised for many years, such as when Benner (1978) noted that:

Practical difficulties arise during the investigation and reporting of most accidents. These difficulties include the determination of the scope of the phenomenon to investigate, the identification of the data required, documentation of the findings, development of recommendations based on the accident findings, and preparation of the deliverables at the end of the investigation. These difficulties reflect differences in the purposes for the investigations, which in turn reflect different perceptions of the accident phenomenon.

The 'different perceptions of the accident phenomenon' are what in present day terminology are called the accident models. Accident models seem to have started by relatively uncomplicated single-factor models of, e.g., accident proneness (Greenwood & Woods, 1919) and developed via simple and complex linear causation models to present-day systemic or functional models (for a recent overview of accident models, see Hollnagel, 2004.)

The archetype of a simple linear model is Heinrich's (1931) Domino model, which explains accidents as the linear propagation of a chain of causes and effects (Figure 1.1). This model was associated with one of the earliest attempts of formulating a complete theory of safety, expressed in terms of ten axioms of industrial safety (Heinrich et al., 1980, p. 21). The first of these axioms reads as follows:

> The occurrence of an injury invariably results from a completed sequence of factors – the last one of these being the accident itself. The accident in turn is invariably caused or permitted directly by the unsafe act of a person and/or a mechanical or physical hazard.

According to this view, an accident is basically a disturbance inflicted on an otherwise stable system. Although the domino model has been highly useful by providing a concrete approach to understanding accidents, it has unfortunately also reinforced the misunderstanding that accidents have a root cause and that this root cause can be found by searching backwards from the event through the chain of causes that preceded it. More importantly, the domino model suggests that system safety can be enhanced by disrupting the linear sequence, either by 'removing' a 'domino' or by 'spacing' the 'dominos' further apart. (The problems in providing a translation from model components to the world of practice are discussed further in Chapter 17.)

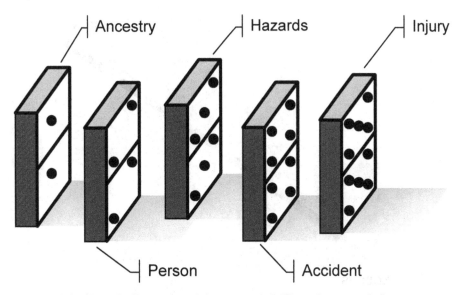

Figure 1.1: Simple linear accident model (Domino model)

The comparable archetype of a complex linear model is the well-known Swiss cheese model first proposed by Reason (1990). According to this, accidents can be seen as the result of interrelations between real time 'unsafe acts' by front-line operators and latent conditions such as weakened barriers and defences, represented by the holes in the slices of 'cheese', cf. Figure 1.2. (Models that describe accidents as a result of interactions among agents, defences and hosts are also known as epidemiological accident models.) Although this model is technically more complex than the domino model, the focus remains on structures or components and the functions associated with these, rather than on the functions of the overall system as such. The Swiss cheese model comprises a number of identifiable components where failures (and risks) are seen as due to failures of the components, most conspicuously as the breakdown of defences. Although causality is no longer a *single* linear propagation of effects, an accident is still the result of a relatively clean combination of events, and the failure of a barrier is still the failure of an individual component. While the whole idea of a complex linear model such as this is to describe how coincidences occur, it cannot detach itself from a structural perspective involving the fixed relations between agents, hosts, barriers and environments.

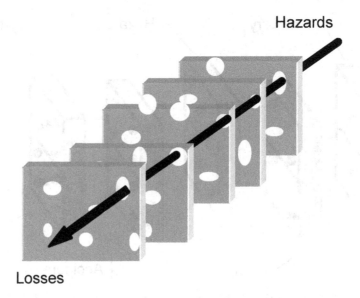

Figure 1.2: Complex linear accident model (Swiss cheese model)

Since some accidents defy the explanatory power of even complex linear models, alternative explanations are needed. Many authors have pointed out that accidents can be seen as due to an unexpected combination or aggregation of conditions or events (e.g., Perrow, 1984). A practical term for this is *concurrence*, meaning the temporal property of two (or more) things happening at the same time and thereby affecting each other. This has led to the view of accidents as non-linear phenomena that emerge in a complex system, and the models are therefore often called systemic accident models, cf., Chapter 4.

This view recognises that complex system performance always is variable, both because of the variability of the environment and the variability of the constituent subsystems. The former may appropriately be called exogenous variability, and the latter endogenous variability. The endogenous variability is to a large extent attributable to the people in the system, as individuals and/or groups. This should nevertheless not be taken to imply that human performance is wrong or erroneous in any way. On the contrary, performance variability is necessary if a joint cognitive system, meaning a human-machine system or a socio-technical system, is successfully to cope with the complexity of the real world (Hollnagel & Woods, 2005). The essence of the systemic view can be expressed by the following four points:

- Normal performance and as well as failures are emergent phenomena. Neither can therefore be attributed to or explained by referring to the (mal)functions of specific components or parts. Normal performance furthermore differs from *normative* performance: it is not what is prescribed by rules and regulation but rather what takes place as a result of the adjustments required by a partly unpredictable environment. Technically speaking, normal performance represents the equilibrium that reflects the regularity of the work environment.

- The outcomes of actions may sometimes differ from what was intended, expected or required. When this happens it is more often due to variability of context and conditions than to the failures of actions (or the failure of components or functions). On the level of individual human performance, local optimisation or adjustment is the norm rather than the exception as shown by the numerous shortcuts and heuristics that people rely on in their work.

- The adaptability and flexibility of human work is the reason for its efficiency. Normal actions are successful because people adjust to local conditions, to shortcomings or quirks of technology, and to predictable changes in resources and demands. In particular, people quickly learn correctly to anticipate recurring variations; this enables them to be proactive, hence to save the time otherwise needed to assess a situation.

- The adaptability and flexibility of human work is, however, also the reason for the failures that occur, although it is rarely the cause of such failures. Actions and responses are almost always based on a limited rather than complete analysis of the current conditions, i.e., a trade-off of thoroughness for efficiency. Yet since this is the normal mode of acting, normal actions can, by definition, not be wrong. Failures occur when this adjustment goes awry, but both the actions and the principles of adjustment are technically correct.

Accepting a specific model does not only have consequences for how accidents are understood, but also for how resilience is seen. In a simple linear model, resilience is the same as being impervious to specific causes; using the domino analogy, the pieces either cannot fall or are so far apart that the fall of one cannot affect its neighbours. In a complex linear model, resilience is the ability to maintain effective barriers that can withstand the impact of harmful agents and the erosion that is a result of latent conditions. In both cases the transition from a safe to an unsafe state is tantamount to the failure of some component or subsystem and resilience is the ability to endure harmful influences. In contrast to that, a systemic model adopts a functional point of view in which resilience is an organisation's ability efficiently to adjust to harmful influences rather than to shun or resist them. An unsafe state may arise because system adjustments are insufficient or inappropriate rather than because something fails. In this view failure is the flip side of success, and therefore a normal phenomenon.

Anticipating Risks

Going from accident analysis to risk assessment, i.e., from understanding what *has* happened to the identification of events or conditions that in the future *may* endanger system safety, it is also possible to find a number of different models of risks, although the development has been less noticeable. Just as there are single-factor accident models, there are risk assessment models that consider the failure of individual components, such as Failure Mode and Effects Analysis. Going one step further, the basic model to describe a sequence of actions is the event tree, corresponding to the simple linear accident model. The event tree represents a future accident as a result of possible failures in a pre-determined sequence of events organised as a binary branching tree. The 'root' is the initiating event and the 'leaves' are the set of possible outcomes – either successes or failures. In a similar manner, a fault tree corresponds to a complex linear model or an epidemiological model. The fault tree describes the accident as the result of a series of logical combinations of conditions, which are necessary and sufficient to produce the top event, i.e., the accident (cf. Figure 1.3).

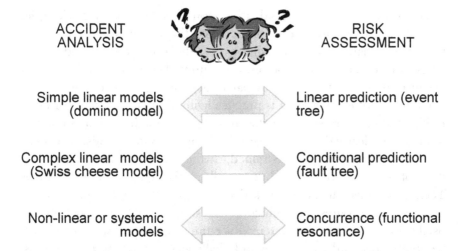

ACCIDENT ANALYSIS		RISK ASSESSMENT
Simple linear models (domino model)	⟺	Linear prediction (event tree)
Complex linear models (Swiss cheese model)	⟺	Conditional prediction (fault tree)
Non-linear or systemic models	⟺	Concurrence (functional resonance)

Figure 1.3: Models for accident analysis and risk assessment

Event trees and fault trees may be adequate for risk assessment when the outcomes range from incidents to smaller accidents, since these often need less elaborate explanations and may be due to relatively simple combinations of factors. Most major accidents, however, are due to complex concurrences of multiple factors, some of which have no apparent *a priori* relations. Event and fault trees are therefore unable fully to describe them – although this does not prevent event trees from being the favourite tool for Probabilistic Safety Assessment methods in general. It is, indeed, a consequence of the systemic view that the potential for (complex) accidents cannot be described by a fixed structure such as a tree, graph or network, but must invoke some way of representing dynamic bindings or couplings, for instance as in the functional resonance accident model (Hollnagel, 2004). Indeed, the problems of risk assessment may to a large degree arise from a reliance on graphical representations, which – as long as they focus on descriptions of links between parts – are unable adequately to account for concurrence and for how a stable system slowly or abruptly may become unstable.

The real challenge for system safety, and therefore also for resilience engineering, is to recognise that complex systems are dynamic and that a state of dynamic stability sometimes may change into a state of dynamic instability. This change may be either abrupt, as in an accident, or slow, as in a gradual erosion of safety margins. Complex systems must perforce be dynamic since they must be able to adjust their performance to the conditions, cf. the four points listed above. These adjustments cannot be pre-programmed or built into the system, because they cannot be anticipated at the time of design – and sometimes not even later. It is practically impossible to design for every little detail or every situation that may arise, something that procedure writers have learned to their dismay. Complex systems must, however, be *dynamically* stable, or constrained, in the sense that the adjustments do not get out of hand but at all times remain under control. Technically this can be expressed by the concept of damping, which denotes the progressive reduction or suppression of deviations or oscillation in a device or system (over time). A system must obviously be able to respond to changes and challenges, but the responses must not lead the system to lose control. The essence of resilience is therefore the intrinsic ability of an organisation (system) to maintain or regain a dynamically stable state, which allows it to continue operations after a major mishap and/or in the presence of a continuous stress.

Dictionaries commonly define resilience as the ability to 'recover quickly from illness, change, or misfortune', one suggestive synonym being buoyancy or a bouncing quality. Using this definition, it stands to reason that it is easier to recover from a potentially destabilising disturbance if it is detected early. The earlier an adjustment is made, the smaller the resulting adjustments are likely to be. This has another beneficial effect, which is a corollary of the substitution myth. According to this, artefacts are value neutral in the sense that the introduction of an artefact into a system only has the intended and no unintended effects (Hollnagel & Woods, 2005, p. 101). Making a response to or recovering from a disturbance requires an adjustment, hence a change to the system. Any such change may have consequences that go beyond the local and intended effects. If the consequences from the recovery are small, i.e., if they can effectively be confined to a subsystem, then the likelihood of negative side-effects is reduced and the resilience is therefore higher. As a result of this, the definition of resilience can be modified to be the ability of a system or an organisation to react to and recover from disturbances at an early stage, with minimal effect on the dynamic stability. The challenges to system safety come from instability, and resilience engineering is an expression of the methods and principles that prevent this from taking place.

For the analytical part, resilience engineering amounts to a systemic accident model as outlined above. Rather than looking for causes we should look for concurrences, and rather than seeing concurrences as exceptions we should see them as normal and therefore also as inevitable. This may at times lead to the conclusion that even though an accident happened nothing really went wrong, in the sense that nothing happened that was out of the ordinary. Instead it is the concurrence of a number of events, just on the border of the ordinary, that constitutes an explanation of the accident or event.

For the predictive part, resilience engineering can be addressed, e.g., by means of a functional risk identification method, such as proposed by the functional resonance accident model (Hollnagel, 2004). To make progress on resilience engineering we have to go beyond failure modes of component (subsystem, human) to concurrences. This underlines the functional view since concurrences take place among events and functions rather than among components. The challenge is understand when a system may lose its dynamic stability and become unstable. To do so requires powerful methods combined with plenty of imagination (Adamski & Westrum, 2003).

Systems are Ever-Changing

Yushi Fujita

No system (i.e., combination of artifact and humans) can avoid changes. Changes occur continuously throughout a system's lifetime. This should be regarded as a destiny. The incompleteness of a system is partly attributable to this ever-changing nature. Changes take place because of external drivers (e.g., economic pressure). But changes also take place because of internal drivers. For instance, humans are always motivated to make changes that they think will improve system administration; humans often find unintended ways of utilizing the artifact; leaders are encouraged to introduce new visions in order to stimulate and lead people; ... Like these, the system is always subject to changes, hence metamorphosing itself like a living matter. This floating nature often causes mismatches between administrative frameworks and the ways in which the system is actually utilized.

Chapter 2

Essential Characteristics of Resilience

David D. Woods

Avoiding the Error of the Third Kind

When one uses the label 'resilience,' the first reaction is to think of resilience as if it were adaptability, i.e., as the ability to absorb or adapt to disturbance, disruption and change. But all systems adapt (though sometimes these processes can be quite slow and difficult to discern) so resilience cannot simply be the adaptive capacity of a system. I want to reserve resilience to refer to the broader capability – how well can a system handle disruptions and variations that fall outside of the base mechanisms/model for being adaptive as defined in that system.

This depends on a distinction between understanding how a system is competent at designed-for-uncertainties, which defines a 'textbook' performance envelope and how a system recognizes when situations challenge or fall outside that envelope – unanticipated variability or perturbations (see parallel analyses in Woods et al., 1990 and Carlson & Doyle, 2000; Csete & Doyle, 2002). Most discussions of definitions of 'robustness' in adaptive systems debate whether resilience refers to first or second order adaptability (Jen, 2003). In the end, the debates tend to settle on emphasizing the system's ability to handle events that fall outside its design envelope and debate what is a design envelope, what events challenge or fall outside that envelope, and how does a system see what it has failed to build into its design (e.g., see url: http://discuss.santafe.edu/robustness/).

The area of textbook competence is in effect a model of variability/uncertainty and a model of how the strategies/plans

/countermeasures in play handle these, mostly successfully. Unanticipated perturbations arise (a) because the model implicit and explicit in the competence envelope is incomplete, limited or wrong and (b) because the environment changes so that new demands, pressures, and vulnerabilities arise that undermine the effectiveness of the competence measures in play.

Resilience then concerns the ability to recognize and adapt to handle unanticipated perturbations that call into question the model of competence, and demand a shift of processes, strategies and coordination. When evidence of holes in the organization's model builds up, the risk is what Ian Mitroff called many years ago, the error of the third kind, or solving the wrong problem (Mitroff, 1974). This is a kind of under-adaptation failure where people persist in applying textbook plans and activities in the face of evidence of changing circumstances that demand a qualitative shift in assessment, priorities, or response strategy.

This means resilience is concerned with monitoring the boundary conditions of the current model for competence (how strategies are matched to demands) and adjusting or expanding that model to better accommodate changing demands. The focus is on assessing the organization's adaptive capacity relative to challenges to that capacity – what sustains or erodes the organization's adaptive capacities? Is it degrading or lower than the changing demands of its environment? What dynamics challenge or go beyond the boundaries of the competence envelope? Is the organization as well adapted as it thinks it is? Note that boundaries are properties of the model that defines the textbook competence envelope relative to the uncertainties and perturbations it is designed for (Rasmussen, 1990a). Hence, resilience engineering devotes effort to make observable the organization's model of how it creates safety, in order to see when the model is in need of revision.

To do this, Resilience Engineering must monitor organizational decision-making to assess the risk that the organization is operating nearer to safety boundaries than it realizes (Woods, 2005a). Monitoring resilience should lead to interventions to manage and adjust the adaptive capacity as the system faces new forms of variation and challenges.

Monitoring and managing resilience, or its absence, brittleness, is concerned with understanding how the system adapts and to what kinds of disturbances in the environment, including properties such as:

- buffering capacity: the size or kinds of disruptions the system can absorb or adapt to without a fundamental breakdown in performance or in the system's structure;
- flexibility versus stiffness: the system's ability to restructure itself in response to external changes or pressures;
- margin: how closely or how precarious the system is currently operating relative to one or another kind of performance boundary;
- tolerance: how a system behaves near a boundary – whether the system gracefully degrades as stress/pressure increase or collapses quickly when pressure exceeds adaptive capacity.

In addition, cross-scale interactions are critical, as the resilience of a system defined at one scale depends on influences from scales above and below:

- Downward, resilience is affected by how organizational context creates or facilitates resolution of pressures/goal conflicts/dilemmas, for example, mismanaging goal conflicts or poor automation design can create authority-responsibility double binds for operational personnel (Woods et al., 1994; Woods, 2005b).
- Upward, resilience is affected by how adaptations by local actors in the form of workarounds or innovative tactics reverberate and influence more strategic goals and interactions (e.g., workload bottlenecks at the operational scale can lead to practitioner workarounds that make management's attempts to command compliance with broad standards unworkable; Cook et al., 2000).

As illustrated in the cases of resilience or brittleness described or referred to in this book, all systems have some degree of resilience and sources for resilience. Even cases with negative outcomes, when seen as breakdowns in adaptation, reveal the complicating dynamics that stress the textbook envelope and the often hidden sources of resilience used to cope with these complexities.

Accidents have been noted by many analysts as 'fundamentally surprising' events because they call into question the organization's model of the risks they face and the effectiveness of the countermeasure deployed (Lanir, 1986; Woods et al., 1994, chapter 5; Rochlin, 1999; Woods, 2005b). In other words, the organization is unable to recognize or interpret evidence of new vulnerabilities or ineffective countermeasures until a visible accident occurs. At this stage the organization can engage in fundamental learning but this window of opportunity comes at a high price and is fragile given the consequences of the harm and losses. The shift demanded following an accident is a reframing process. In reframing one notices initial signs that call into question ongoing models, plans and routines, and begins processes of inquiry to test if revision is warranted (Klein et al., 2005). Resilience Engineering aims to provide support for the cognitive processes of reframing an organization's model of how safety is created before accidents occur by developing measures and indicators of contributors to resilience such as the properties of buffers, flexibility, precariousness, and tolerance and patterns of interactions across scales such as responsibility-authority double binds.

Monitoring resilience is monitoring for the changing boundary conditions of the textbook competence envelope – how a system is competent at handling designed-for-uncertainties – to recognize forms of unanticipated perturbations – dynamics that challenge or go beyond the envelope. This is a kind of broadening check that identifies when the organization needs to learn and change. Resilience engineering needs to identify the classes of dynamics that undermine resilience and result in organizations that act riskier than they realize. This chapter focuses on dynamics related to safety-production goal conflicts.

Coping with Pressure to be Faster, Better, Cheaper

Consider recent NASA experience, in particular, the consequences of NASA's adoption of a policy called 'faster, better, cheaper' (FBC). Several years later a series of mishaps in space science missions rocked the organization and called into question that policy. In a remarkable 'organizational accident' report, an independent team investigated the organizational factors that spawned the set of mishaps (Spear, 2000).

The investigation realized that FBC was not a policy choice, but the acknowledgement that the organization was under fundamental pressure from stakeholders. The report and the follow-up, but short-lived, 'Design for Safety' program noted that NASA had to cope with a changing environment with increasing performance demands combined with reduced resources: drive down the cost of launches, meet shorter, more aggressive mission schedules, do work in a new organizational structure that required people to shift roles and coordinate with new partners, eroding levels of personnel experience and skills. Plus, all of these changes were occurring against a backdrop of heightened public and congressional interest that threatened the viability of the space program. The MCO investigation board concluded: NASA, which had a history of 'successfully carrying out some of the most challenging and complex engineering tasks ever faced by this nation,' was being asked to 'sustain this level of success while continually cutting costs, personnel and development time ... these demands have stressed the system to the limit' due to 'insufficient time to reflect on unintended consequences of day-to-day decisions, insufficient time and workforce available to provide the levels of checks and balances normally found, breakdowns in inter-group communications, too much emphasis on cost and schedule reduction.' The MCO Board diagnosed the mishaps as indicators of an increasingly brittle system as production pressure eroded sources of resilience and led to decisions that were riskier than anyone wanted or realized. Given this diagnosis, the Board went on to re-conceptualize the issue as how to provide tools for proactively monitoring and managing project risk throughout a project life-cycle and how to use these tools to balance safety with the pressure to be faster, better, cheaper.

The experience of NASA under FBC is an example of the law of stretched systems: every system is stretched to operate at its capacity; as soon as there is some improvement, for example in the form of new technology, it will be exploited to achieve a new intensity and tempo of activity (Woods, 2003). Under pressure from performance and efficiency demands (FBC pressure), advances are consumed to ask operational personnel 'to do more, do it faster or do it in more complex ways', as the Mars Climate Orbiter Mishap Investigation Board report determined. With or without cheerleading from prestigious groups, pressures to be 'faster, better, cheaper' increase. Furthermore, pressures to be 'faster, better, cheaper' introduce changes, some of which are new capabilities (the term does include 'better'), and these changes modify the vulnerabilities or paths toward failure. How conflicts and trade-offs like these are recognized and handled in the context of vectors of change is an important aspect of managing resilience.

Balancing Acute and Chronic Goals

Problems in the US healthcare delivery system provide another informative case where faster, better, cheaper pressures conflict with safety and other chronic goals. The Institute of Medicine in a calculated strategy to guide national improvements in health care delivery conducted a series of assessments. One of these, Crossing the Quality Chasm: A New Health System for the 21st Century (IOM, 2001), stated six goals needed to be achieved simultaneously: the national health care system should be – Safe, Effective, Patient-centered, Timely, Efficient, Equitable.[1] Each goal is worthy and generates thunderous agreement. The next step seems quite direct and obvious – how to identify and implement quick steps to advance each goal (the classic search for so-called 'low hanging fruit'). But as in the NASA case, this set of goals is not a new policy direction but rather an acknowledgement of demanding pressures already operating on health care practitioners and organizations. Even more difficult, the six goals represent a set of interacting and often conflicting pressures so that in adapting to reach

---

[1]   The IOM states the quality goals as –
'Health Care Should Be:
*   Safe – avoiding injuries to patients from the care that is intended to help them.
*   Effective – providing services based on scientific knowledge to all who could benefit and refraining from providing services to those not likely to benefit (avoiding underuse and overuse, respectively).
*   Patient-centered – providing care that is respectful of and responsive to individual patient preferences, needs, and values and ensuring that patient values guide all clinical decisions.
*   Timely – reducing waits and sometimes harmful delays for both those who receive and those who give care.
*   Efficient – avoiding waste, including waste of equipment, supplies, ideas, and energy.
*   Equitable – providing care that does not vary in quality because of personal characteristics such as gender, ethnicity, geographic location, and socioeconomic status.'

for one of these goals it is very easy to undermine or squeeze others. To improve on all simultaneously is quite tricky.

As I have worked on safety in health care, I hear many highly placed voices for change express a basic belief that these six goals can be synergistic. Their agenda is to energize a search for and adoption of specific mechanisms that simultaneously advance multiple goals within the six and that do not conflict with others – 'silver bullets'. For example, much of the patient safety discussion in US health care continues to be a search for specific mechanisms that appear to simultaneously save money and reduce injuries as a result of care. Similarly, NASA senior leaders thought that including 'better' along with faster and cheaper meant that techniques were available to achieve progress on being faster, better, and cheaper together (for almost comic rationalizations of 'faster, better, cheaper' following the series of Mars science mission mishaps and an attempt to protect the reputation of the NASA administrator at the time, see Spear, 2000). The IOM and NASA senior management believed that quality improvements began with the search for these 'silver bullet' mechanisms (sometimes called 'best practices' in health care). Once such practices are identified, the question becomes how to get practitioners and organizations to adopt these practices. Other fields can help provide the means to develop and document new best practices by describing successes from other industries (health care frequently uses aviation and space efforts to justify similar programs in health care organizations). The IOM in particular has had a public strategy to generate this set of silver bullet practices and accompanying justifications (like creating a quality catalog) and then pressure health care delivery decision makers to adopt them all in the firm belief that, as a result, all six goals will be advanced simultaneously and all stakeholders and participants will benefit (one example is computerized physician order entry).

However, the findings of the Columbia accident investigation board (CAIB) report should reveal to all that the silver bullet strategy is a mirage. The heart of the matter is not silver bullets that eliminate conflicts across goals, but developing new mechanisms that balance the inherent tensions and trade-offs across these goals (Woods et al., 1994). The general trade-off occurs between the family of acute goals – timely, efficient, effective (or after NASA's policy, the Faster, Better, Cheaper or FBC goals) and the family of chronic goals, for the health care case consisting of safety, patient-centeredness, and equitable access.

The tension between acute production goals and chronic safety risks is seen dramatically in the Columbia accident which the investigation board found was the result of pressure on acute goals eroding attention, energy and investments on chronic goals related to controlling safety risks (Gehman, 2003). Hollnagel (2004, p. 160) compactly captured the tension between the two sets of goals with the comment that:

> If anything is unreasonable, it is the requirement to be both efficient and thorough at the same time – or rather to be thorough when with hindsight it was wrong to be efficient.

The FBC goal set is acute in the sense that they happen in the short term and can be assessed through pointed data collection that aggregates element counts (shorter hospitals stays, delay times). Note that 'better' is in this set, though better in this family means increasing capabilities in a focused or narrow way, e.g., cardiac patients are treated more consistently with a standard protocol. The development of new therapies and diagnostic capabilities belongs in the acute sense of 'better.'

Safety, access, patient-centeredness are chronic goals in the sense that they are system properties that emerge from the interaction of elements in the system and play out over longer time frames. For example, safety is an emergent system property, arising in the interactions across components, subsystems, software, organizations, and human behavior.

By focusing on the tensions across the two sets, we can better see the current situation in health care. It seems to be lurching from crisis to crisis as efforts to improve or respond in one area are accompanied by new tensions at the intersections of other goals (or the tensions are there all along and the visible crisis point shifts as stakeholders and the press shift their attention to different manifestations of the underlying conflicts). The tensions and trade-offs are seen when improvements or investments in one area contribute to greater squeezes in another area. The conflicts are stirred by the changing background of capabilities and economic pressure. The shifting points of crisis can be seen first in 1995-6 as dramatic well publicized deaths due to care helped create the patient safety crisis (ultimately documented in Kohn et al., 1999). The patient safety movement was energized by patients feeling vulnerable as health care changed to meet cost control pressures. Today attention has shifted to an access crisis as malpractice rates and prescription drug costs undermine patients' access to physicians in high risk specialties and challenge seniors' ability to balance medication costs with limited personal budgets.

Dynamic Balancing Acts

If the tension view is correct, then progress revolves around how to dynamically balance the potential trade-offs so that all six goals can advance (as opposed to the current situation where improvements or investments in one area create greater squeezes in another area). It is important to remember that trade-offs are defined by two parameters, one that captures discrimination power or how well one can make the underlying judgement, and a second that defines where to place a criterion for making a decision or taking an action along the trade-off curve, criterion placement or movement. The parameters of a trade-off cannot be estimated by a single case, but require integration over behavior in sets of cases and over time.

One aspect of the difficulty of goal conflicts is that the default or typical ways to advance the acute goals often make it harder to achieve chronic goals simultaneously. For example, increasing therapeutic capabilities can easily appear as new silos of care that do not redress and can even exacerbate fragmentation of care (undermining the patient-centeredness goal). To advance all of the goals, ironically, the chronic set of goals of patient centered, safety and access must be put first, with secondary concern for efficient and timely methods. To do otherwise will fall prey to the natural tendency to value the more immediate and direct consequences (which, by the way, are easier to measure) of the acute set over the chronic and produce an unintentional sacrifice on the chronic set. Effective balance seems to arise when organizations shift from seeing safety as one of a set of goals to be measured (is it going up or down?) to considering safety as a basic value. The point is that for chronic goals to be given enough weight in the interaction with acute goals, the chronic needs to be approached much more like establishing a core cultural value.

For example, valuing the chronic set in health care puts patient centeredness first with its fellow travelers safety and access. The central issue under patient centeredness is emergent continuity of care, as the patient makes different encounters with the health care system and as disease processes develop over time. The opposite of continuity is fragmentation. Many of the tensions across goals exacerbate fragmentation, e.g., ironically, new capabilities on specific aspects of health care can lead to more specialization and more silos of care. Placing priority on continuity of care vs. fragmentation focuses attention (a) on health care issues related to chronic diseases which require continuity and which are inherently difficult in a fragmented system of care and (b) on cognitive system issues which address coordination over time, over practitioners, over organizations, and over specialized knowledge sources. Consider the different ways new technology can have an effect on patient care. Depending on how computer systems are built and adapted over time, more computerization can lead to less contact with patients and more contact with the image of the patient in the database. This is a likely outcome when FBC pressure leads acute goals to dominate chronic ones (the benefits of the advance in information technology will tend to be consumed to meet pressures for productivity or efficiency). When a chronic goal such as continuity of care, functions as the leading value, the emphasis shifts to finding uses of computer capabilities that increase attention and tailoring of general standards to a specific patient over time (increasing the effective continuity) and only then developing these capabilities to meet cost considerations.

The tension diagnosis is part of the more general diagnosis that past success has led to increasingly complex systems with new forms of problems and failure risks. The basic issue for organizational design is how large-scale systems can cope with complexity, especially the pace of change and coupling across parts that accompany the methods that advance the acute goals. To miss the complexity diagnosis will make otherwise well-intentioned efforts fail as each attempt to advance goals simultaneously through silver bullets will rebound as new crises where goal trade-offs create new dissatisfactions and tensions.

Sacrifice Judgements

To illustrate a safety culture, leaders tell stories about an individual making tough decisions when goals conflict. The stories always have the same basic form even though the details may come from a personal experience or from re-telling of a story gathered from another domain with a high reputation for safety (e.g., health care leaders often use aerospace stories):

> Someone noticed there might be a problem developing, but the evidence is subtle or ambiguous. This person has the courage to speak up and stop the production process underway. After the aircraft gets back on the ground or after the system is dismantled or after the hint is chased down with additional data, then all discover the courageous voice was correct. There was a problem that would otherwise have been missed and to have continued would have resulted in failure, losses, and injuries. The story closes with an image of accolades for the courageous voice.

> When the speaker finishes the story, the audience sighs with appreciation – that was an admirable voice and it illustrates how a great organization encourages people to speak up about potential safety problems. You can almost see people in the audience thinking, 'I wish my organization had a culture that helped people act this way.'

> But this common story line has the wrong ending. It is a quite different ending that provides the true test for a high resilience organization.

> When they go look, after the landing or after dismantling the device or after the extra tests were run, everything turns out to be OK. The evidence of a problem isn't there or may be ambiguous; production apparently did not need to be stopped. Now, how does the organization's management react? How do the courageous voice's peers react?

> For there to be high resilience, the organization has to recognize the voice as courageous and valuable even though the result was apparently an unnecessary sacrifice on production and efficiency goals. Otherwise, people balancing multiple goals will tend to act riskier than we want them to, or riskier than they themselves really want to.

These contrasting story lines illustrate the difficulties of balancing acute goals with chronic ones. Given a backdrop of schedule pressure, how should an organization react to potential 'warning' signs and seek to handle the issues the signs point to? If organizations never sacrifice production pressure to follow up warning signs, they are acting much too risky. On the other hand, if uncertain 'warning' signs always lead to sacrifices on acute goals, can the organization operate within reasonable parameters or stakeholder demands? It is easy for organizations that are working hard to advance the acute goal set to see such warning signs as risking inefficiencies or as low probability of concern as they point to a record of apparent success and improvement. Ironically, these same signs after-the-fact of an accident appear to all as clear cut undeniable warning signs of imminent dangers.

To proactively manage risk prior to outcome requires ways to know when to relax the pressure on throughput and efficiency goals, i.e., making a sacrifice judgement. Resilience engineering needs to provide organizations with help on how to decide when to relax production pressure to reduce risk (Woods, 2000). I refer to these trade-off decisions as sacrifice judgements because acute production or efficiency related goals are temporarily sacrificed, or the pressure to achieve these goals is relaxed, in order to reduce the risks of approaching too near safety boundaries. Sacrifice judgements occur in many settings: when to convert from laparoscopic surgery to an open procedure (Dominguez et al., 2004 and the discussion in Cook et al., 1998), when to break off an approach to an airport during weather that increases the risks of wind shear, and when to have a local slowdown in production operations to avoid risks as complications build up.

New research is needed to understand this judgement process in individuals and in organizations. Previous research on such decisions (e.g., production/safety trade-off decisions in laparoscopic surgery) indicates that the decision to value production over safety is implicit and unrecognized. The result is that individuals and organizations act much riskier than they would ever desire. A sacrifice judgement is especially difficult because the hindsight view will indicate that the sacrifice or relaxation may have been unnecessary since 'nothing happened.' This means that it is important to assess how peers and superiors react to such decisions.

The goal is to develop explicit guidance on how to help people make the relaxation/sacrifice judgement under uncertainty, to maintain a desired level of risk acceptance/risk averseness, and to recognize changing levels of risk acceptance/risk averseness. For example, what indicators reveal a safety/production trade-off sliding out of balance as pressure rises to achieve acute production and efficiency goals? Ironically, it is these very times of higher organizational tempo and focus on acute goals that require extra investments in sources of resilience to keep production/safety trade-offs in balance – valuing thoroughness despite the potential for sacrifices on efficiency required to meet stakeholder demands.

Note how the recommendation to aid sacrifice judgements is a specialization of general methods for aiding any system confronting a trade-off: (a) improve the discrimination power of the system confronting the trade-off, and (b) help the system dynamically match its placement of a decision criterion with the assessment of changing risk and uncertainty.

Resilience Engineering should provide the means for dynamically adjusting the balance across the sets of acute and chronic goals. The dilemma of production pressure/safety trade-offs is that we need to pay the most attention to, and devote scarce resources to, potential future safety risks when they are least affordable due to increasing pressures to produce or economize. As a result, organizations unknowingly act riskier than they would normally accept. The first step is tools to monitor the boundary between competence at designed-for-uncertainties and unanticipated perturbations that challenge or fall outside that envelope. Recognizing signs of unanticipated perturbations consuming or stretching the sources of resilience in the system can lead actions to re-charge a system's resilience. How can we increase, maintain, or re-establish resilience when buffers are being depleted, margins are precarious, processes become stiff, and squeezes become tighter?

Acknowledgements

This work was supported in part by grant NNA04CK45A from NASA Ames Research Center to develop resilience engineering concepts for managing organizational risk. The ideas presented benefited from discussions in the NASA's Design for Safety workshop and Workshop on organizational risk. Discussions with John Wreathall helped develop the model of trade-offs across acute and chronic goals.

Chapter 3

Defining Resilience

Andrew Hale
Tom Heijer

Pictures of Resilience

Resilience first conjures up in the mind pictures of bouncing back from adversity: the boxer dancing courageously out of his corner despite the battering in the previous round; the kidnap victim, like Terry Waite, emerging from months of privation as a prisoner of terrorist organisations, smiling and talking rationally about his experiences; the victim of a severe handicap, like Stephen Hawkins or Christopher Reeve, still managing to make a major contribution to society or a chosen cause. If we were to apply this image to organisations, the emphasis would come to fall on responding to disaster: rapid recovery from a disastrous fire by reopening production in a temporary building; restoring confidence among local residents after a major chemical leak by full openness over the investigation and involvement in decisions about improved prevention measures; or restoring power on the network after a major outage by drafting in extra staff to work around the clock. This captures some of the essentials, with an emphasis on flexibility, coping with unexpected and unplanned situations and responding rapidly to events, with excellent communication and mobilisation of resources to intervene at the critical points. However, we would argue that we should extend the definition a little more broadly, in order to encompass also the ability to avert the disaster or major upset, using these same characteristics. Resilience then describes also the characteristic of managing the organisation's activities to anticipate and circumvent threats to its existence and primary goals.

This is shown in particular in an ability to manage severe pressures and conflicts between safety and the primary production or performance goals of the organisation.

This definition can be projected easily onto the model which Rasmussen proposed to understand the drift to danger (Figure 3.1, adapted from Rasmussen & Svedung, 2000).

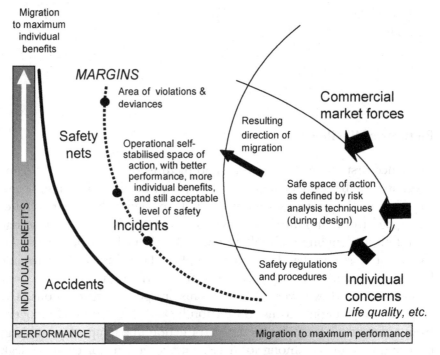

Figure 3.1: Rasmussen's drift to danger model

This aspect of resilience concentrates on the prevention of loss of control over risk, rather than recovery from that loss of control. If we use the representation of the bowtie model of accident scenarios (Figure 3.2 adapted from Visser, 1998), we are locating resilience not only on the right-hand side of the centre event, but also on the left.

Reverting to Rasmussen's model, resilience is the ability to steer the activities of the organisation so that it may sail close to the area where accidents will happen, but always stays out of that dangerous area. This implies a very sensitive awareness of where the organisation is in relation to that danger area and a very rapid and an effective response when signals of approaching or actual danger are detected, even unexpected or unknown ones. The picture this conjures up is of a medieval ship with wakeful lookouts, taking constant soundings as it sails in the unknown and misty waters of the far north or west, alert for icebergs, terrible beasts or the possibility of falling off the edge of the flat earth. We cannot talk of resilience unless the organisation achieves this feat consistently over a long period of time. As the metaphor of the ship implies, resilience is a dynamic process of steering and not a static state of an organisation. It has to be worked at continuously and, like the voyage of the Flying Dutchman, the task is never ended and the resilience can always disappear or be proven ineffective in the face of particular threats.

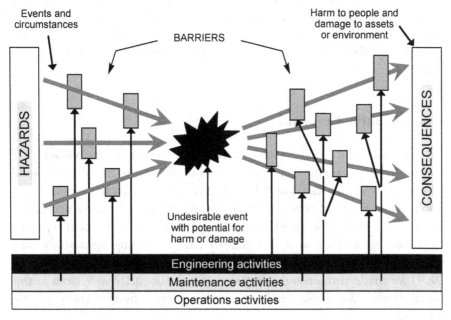

Figure 3.2: Bowtie model

How Do We Recognise Resilience When We See It?

These characteristics of resilience are worked out in more detail elsewhere in this book. However, we wish to add one more aspect of a definition at this point. This is something we need to consider when asking the question: 'how do we recognise a resilient organisation?' One answer to this is to rely on measuring the sort of characteristics which have been set out above. Are they present in the organisation or system? However, this still leaves us with the dilemma as to whether these characteristics are indeed the ones that are crucial in achieving that dynamic avoidance of accidents and disasters. We would really like to have confirmation with an outcome measure over a long period. The obvious one is the safety performance of the organisation or activity over a long period, coupled with its survival (and preferably prosperity) on the basis of other performance measures such as production, service, productivity and quality. The question then arises whether a safe organisation is by definition a resilient one and whether one which has accidents is by definition not resilient. We shall explore the first question in Chapter 9 by looking at railways, but suffice it here to say that the answer is 'no'. Organisations can be safe without being resilient. A few observations about the second question are given below.

Is Road Traffic Resilient?

Let us take the road system and ask whether it is resilient. It undoubtedly has many accidents. It is high up on the list of worldwide killers compiled by the World Health Organisation. In the Netherlands it kills more than ten times as many people as die from work accidents. So it does not look good for its credentials as a resilient system. However, let us look at it in terms of risk. In setting up a study into interaction behaviour between traffic participants (Houtenbos et al., 2004) we made some rough estimates of risk for the road system in the Netherlands. We took 'back of the envelope' figures such as the following:

- 1.3×10^{11} vehicle kilometres per year (cf., www.swov.nl)
- An estimated average of 5 encounters/km with other road users; encounters being meetings with other traffic at road junctions or during overtaking manoeuvres, pedestrians or cyclists crossing the road, etc., where we can consider that a potential for accident exists. (We have taken a range from 1 encounter/km for motorways on which 38% of Dutch vehicle kilometres are driven,

up to 20/km on 50kph roads in urban areas, on which 26% of the kilometres are driven (www.swov.nl), with intervening numbers for 80 and 100kph roads. These are figures we estimated ourselves and have no empirical basis.)

This means that there are some 6.5×10^{11} encounters/year for a death toll of a little over 1000/year. The vast majority of these are the result of an encounter between at least two traffic participants. In a more accurate analysis we would need to subtract one-sided accidents where the vehicle leaves the road due to excessive speed, loss of adhesion, etc. This gives us a risk of death of 1.5×10^{-9}/encounter. We know that there are far more injury and damage accidents than deaths, but even if this is a factor 10,000 more, the accident rate per encounter is still only 1.5×10^{-5}. This is extremely low, particularly when we consider that typical figures used in quantitative risk assessment to estimate the probability of human error in even simple tasks are never below 10^{-4}. Is the road system resilient on the basis of this argument? Given the exposure to situations which could lead to accidents, we would argue that it is, despite the accident toll. This led us to conclude that the participants manage these encounters very effectively using local clues and local interactions, and to decide that we needed to study much more thoroughly how the encounters are handled before jumping to all sorts of conclusions about the need to introduce complex information technology to improve the safety of these interactions in the system.

In the other direction, we can ask whether one of the reasons why the aviation system has such a good safety performance – particularly in the space between take-off and landing – is simply that airspace, certainly compared with the roads in the Netherlands, is amazingly empty and so encounters, even without air traffic management (ATM) control, would be very rare? By this we do not mean to call into question whether aviation is ultra-safe or not, but to put the question whether safety performance alone can be seen as proof of resilience without taking into account the issue of exposure to risk. We do this particularly because we see some similar (potential) mechanisms in both systems, which lead to high safety performance, and which we would see as characteristic of resilience. Notable among these is the (potential) presence of simple local rules governing interactions, without the need for central controlling intervention. Hoekstra (2001) demonstrated with simulations just how successful such local rules for governing encounters under 'free flight' rules (with no ATM control) could be. We see just such 'rules' being used by drivers on the road at intersections (Houtenbos et al., 2004) and they also seem to work amazingly well most of the time.

Conclusion

In this short note we are pleading for two things: what is interesting for safety is preventing accidents and not just surviving them. If resilience is used with its common meaning of survival in adversity, we do not see it to be of interest to us. If its definition is extended to cover the ability in difficult conditions to stay within the safe envelope and avoid accidents it becomes a useful term. We would, however, ask whether we do not have other terms already for that phenomenon, such as high reliability organisations, or organisations with an excellent safety culture.

We would enter a plea that we should consider resilience against the background of the size of the risk. You can be resilient in situations of very high risk and still have a quite substantial number of accidents. The significant thing is that there are not far more. You can also fail to be resilient in conditions of low risk and have no accidents. We will argue in Chapter 9 that you can also perform well in major risk conditions without being resilient.

Nature of Changes in Systems

Yushi Fujita

Humans in systems (e.g., operators, maintenance people) are essentially alike and are, in general, adaptive and proactive. These are admirable qualities, but of limited scope. Adaptive and proactive behaviors can change systems continuously, but humans at the front end alone may or may not be able to recognise the potential impact that the changes can have on the system, especially the impact when several changes are put into effect simultaneously. Humans at the back end (e.g., administrators, regulators) tend to have sanguine ideas such as that the system is always operated as planned, and that rules and procedures can fix the system at an optimal state. Mismatches caused by these two tendencies constitute latent hazards, which may cause the system to drift to failures.

Chapter 4

Complexity, Emergence, Resilience …

Jean Pariès

Future is inevitable, but it may not happen
Jorges Luis Borges

Introduction

If safety is simply taken as the ability to operate without killing (too many) people, organizational resilience does not necessarily mean safety. If resilience is be taken as the intrinsic capacity of an organization to recover a stable state (the initial one or a new one) allowing it to continue operations after a major mishap or in presence of a continuous stress, then the ability of an organization to ensure its own survival/operations against adverse circumstances may well imply being momentarily unsafe for its members, or other stakeholders. A good example of that is a country in war, defending itself against military aggression. Resilience may imply accepting to lose some 'lives' among the people. However, from a different perspective, which is widely developed in this book, safety can also be seen as a form of resilience, i.e., as the result of the robustness of all the processes that keep the system safe towards all kinds of stressors, pathogens, threats and the like.

A further strong contention of this book is that organizational resilience is an emerging property of complex systems. The goal of this short section is to elaborate on the notion of emergence. It will not discuss the notion of resilience itself, extensively addressed elsewhere throughout this book. The focus will rather be put on the relationship between complexity and emergence. The issues that will be addressed are questions such as what does it mean that resilience (or any other property) is an emerging property, what the relationship is between emergence and complexity, and whether a more complex system necessarily is less resilient, as Perrow (1984) suggested. The goal is to provide the reader of this book with a framework and some insight (and hopefully some questions left) about these transverse notions, in order to facilitate the comprehension of the discussions in the other chapters about resilience and the challenge of engineering it into organizations.

Emergence and Systems

In the Middle Age in the alchemists' laboratories, persistent researchers were seeking the long life elixir – the infallible remedy for all diseases and the philosophers' stone – able to trigger the transmutation of common metal in gold (Figure 4.1). While they did not really succeed, they did discover a lot of less ambitious properties of matter, and above all, they realised that mixtures could have properties that components had not. In other words, they discovered the power of interaction. And progressively, alchemists became chemists.

Since then, science has been reductionistic. Scientists have been decomposing phenomena, systems and matter into interacting parts, explaining properties at one level from laws describing the interaction of component properties at a lower level of organization. And because more and more components (e.g., atoms) are shared by all phenomena as we go deeper into the decomposition, science has been able to reduce complexity. A multitude of phenomena are explained by a few laws and by the particular parameters of the case. Conversely, as in chess where a small number of rules can generate a huge number of board configurations, a few laws and parameters generate the diversity of properties and phenomena. One could say that properties emerge from the interaction of lower level components. As Holland (1998) puts it, 'emergence in rule-governed systems comes close to being the obverse of reduction'.

So there is a strong relationship between emergence and complexity, as well as between complexity and explanation. 'Emergence' is what happens when we try to understand the properties of a system that exceeds the level of size and complexity that our intellect can grasp at once, so we decompose the system into interacting component parts. But an obvious question immediately arises from here: can we 'explain' all the properties of the world (physical, biological, psychological, social, etc.) through such a reduction process? Clearly, the least one can say is that the reductionistic strategy does not have the same efficiency for all aspects of the world.

Figure 4.1: The alchemists' laboratory

If I try to explain the weight of my alarm clock from the weight of its constituents, it works pretty well. When I try to explain the capability of my alarm clock to wake me up at a specific time, it is clearly no longer an issue of summing up individual components' properties: 'the whole is greater than the sum of its parts'. But I can still figure out and describe how components' interactions complement each other to wake me up.

Now, if one tried to generate an explanation of my waking-up, from the properties of the atoms composing my body, the challenge might well be simply insuperable. Only a living being can wake up. No atoms of my body are living, yet I am living (Morowitz, 2002). Life, but also societies, economies, ecosystems, organizations, consciousness, have properties that cannot be deducted from their component agents' properties, and even have a rather high degree of autonomy from their parts. Millions of cells of my body die and are replaced every day, but I am still myself.

So why is it that we cannot derive these properties from components' properties? It may be that we could in principle – that every aspect, property, state of the world is entirely understandable in the language of chemistry – notwithstanding our computing limitation, which would make any mathematical and computational approach overwhelmingly long and complex. The famous French mathematician and astronomer Pierre Laplace nicely captured this vision, through his 'demon' metaphor:

> We may regard the present state of the universe as the effect of its past and the cause of its future. An intellect which at any given moment knew all of the forces that animate nature and the mutual positions of the beings that compose it, if this intellect were vast enough to submit the data to analysis, could condense into a single formula the movement of the greatest bodies of the universe and that of the lightest atom; for such an intellect nothing could be uncertain and the future just like the past would be present before its eyes. (Laplace, 1814)

The contention here is that the states of a deterministic macro system, such as a collection of particles, are completely fixed once the laws (e.g., Newtonian mechanics) and the initial/boundary conditions are specified at the microscopic level, whether or not we poor Humans can actually predict these states through computation. This is one form of possible relationship between micro and macro phenomena, in which the causal dynamics at one level are entirely determined by the causal dynamics at lower levels of organization.

The notion of emergence is then simply a broader form of relationship between micro and macro levels, in which properties at a higher level are both dependent on, and autonomous from, underlying processes at lower levels. (Bedau, 1997, 2002) has categorized emergence into three categories.

- Nominal emergence when macro level properties, while meaningless at the micro level, can be derived from assembling micro level properties. Innumerable illustrations of nominal emergence range from physics (e.g., from atomic structure to the physical properties of a metal) to all designed systems (e.g., see the alarm-clock example above);

- Weak emergence when an 'in principle' microscopic account of macroscopic behaviour is still possible, but the detailed and comprehensive behaviour cannot be predicted without performing a one-to-one simulation, because there is no condensed explanation[1] of the system's causal dynamics. Illustrations of weak emergence are provided by insect swarms, neural networks, cellular automata, and the like.

- Strong emergence when macro-level properties cannot be explained, and even less predicted, even in principle, by any micro-level causality. Strong emergence would defeat even Laplace's unlimited computer demon. The existence of strong emergence is a contentious issue, as it is inconsistent with the common scientific dogma of upward causation, and introduces a presumption that holistic, downward causation binds the underlying laws of physics to impose organizing principles over components. For a very exciting contribution to this discussion, for cosmology as well as for biology, see Davies (2004). He makes the point that a system above a specific level of complexity (of which the order of magnitude can be computed) cannot be entirely controlled by upward causation, because of the existence of fundamental upper bounds on information content and the information processing rate.

From Emergence to Resilience

The above discussion about different types of emergence may provide some insight into the notion of resilience. If, as discussed elsewhere in

---

[1] The complexity of an object can be expressed by the Kolmogorov measure, which is the length of the shortest algorithm capable of generating this object. In the case of weak emergence, the system dynamics are algorithmically incompressible in the Kolmogorov sense. In this case, the fastest simulator of the system is the system itself.

this book, organizational resilience is an emergent property of complex organizations, one could build on the above-mentioned taxonomy of emergence and define nominal, weak and strong organizational resilience. Among others, an expected benefit of doing that could be to get a broader perspective of the respective contributions of the 'sharp end' and the 'blunt end' in the aetiology of accidents. A question is whether the focus on 'organizational psychopathology' has been overplayed during the last few years' effort to go beyond the front-line operators' error perspective in the understanding of accidents, and whether 'we should redress some of our efforts back to the human at the sharp end' (Shorrock et al., 2005). In his book *Managing the risks of organizational accidents*, J. Reason (1997) warned that 'the pendulum may have swung too far in our present attempts to track down possible errors and accident contributions that are widely separated in both time and place from the events themselves'. Differentiating the construction of organizational resilience according to the nature of its underlying emergence process may be a way to reconcile the sharp end and blunt end perspectives into a better integrated vision.

So what would 'nominal', 'weak', and 'strong' organizational resilience stand for? Inspired by the discussion of the previous section, one could suggest definitions as follows.

- Nominally Emergent Resilience (NER) would refer to resilience features of an organization resulting from a 'computable' combination of individual agent properties. These may well themselves be complex emergent properties, such as consciousness, risk awareness, etc. NER would include all what contributes to the reliability, and also to the resilience, of local couplings between individual agents and their environment. It would cover individual human features (e.g., individual cognition, error management, surprise management, stress management), as well as organizational measures capable of efficacy over these features (e.g., the design of error tolerant environments). It would also include the interactions between individual agents that can be rationally perceived as facilitating collective resilience at different scales, from small groups to large communities: role definition, shared procedures, communication, leadership, delegation, cross-monitoring, and the like. Finally, it would include most aspects of a Safety Management System.

- Weakly Emergent Resilience (WER) would refer to the resilience features of an organization resulting from a combination of individual agent properties in such a way that, although in principle 'computable', the detailed and comprehensive resilient macroscopic behaviour cannot be predicted without performing a one-to-one simulation. This will happen with complex self-regulated systems in which feedback and feedforward loops interact to govern the behaviour of the system (Leveson, 2004). This will also typically happen for some, although not for all, large scale (large population) systems. Indeed, such systems can remain 'simple' if mutual relationships between basic components tend to annihilate individual deviations: individual variety is then submerged by statistical mean values, so that microscopic complexity disappears at macroscopic scales (e.g., Boltzmann's statistical interpretation of entropy in physics). Large scale systems may also become 'complex' when individual variety at microscopic levels is combined and amplified so that it creates new properties at macroscopic level. Simplified illustrations of these phenomena are provided by insect colonies ('smart swarms'), (see box below, also Bonabeau et al., 1999, 2000), neural networks, collective robotics intelligence (Kube & Bonabeau, 2000), and the like. A key mechanism here is the indirect coupling between component agents through modifications that their individual behaviour introduces into the shared environment.

Individual ants run random trips, and put down a chemical trace (pheromone) on their way (Figure 4.2). Pheromone works as a probabilistic guidance for ants: they tend to follow it, but often lose the track. When one ant comes across a prey, it will automatically bite it and follow its own trace, hence heading back to the nest. As it still puts down pheromone, it will also reinforce the chemical track, increasing the chances for another ant to follow the track to the prey. From a small set of very simple individual behavioural rules ('follow pheromone', 'bite preys'), a self-organizing, auto-catalytic process is thus triggered, amplifying guidance to the prey, while individual 'errors' of ants losing the track allow for new prey discovery. A memory of positive past experience is literally written into the environment, leading to an efficient and adaptive collective strategy to collect food. A 'collective teleonomy' seems to emerge, while no direct communication has taken place, and no representation of the task whatsoever is accessible to individuals (after Bonabeau et al., 1999; 2000).

Ants have virtually no brains (about one hundred neurons, versus about one hundred billion for humans). One could then dryly reject the legitimacy of any comparison, even metaphoric, between insect swarms and human societies. Nevertheless, humans do exhibit collective behaviour based on the interaction of rather simple individual rules. Group or crowd behaviour during aircraft, ship, tunnel or building evacuation, or simply daily pedestrian flows or automobile traffic jams can be and have been modelled using software simulators based on low-level individual rules interaction (e.g., Galea, 2001). Similar, but much more complex interactions are behind the notion of distributed cognition (Hutchins, 1996). And from a more philosophical point of view, the anthropologist René Girard (1961) has been able to build a fascinating, although controversial, holistic theory of mankind behaviour (including myths, religions, power, laws, war, and so on) based on the mere idea that humans tend to imitate each other.

Figure 4.2: Ants and smart cooperation

From a resilience perspective, this category of complex systems has interesting properties. First, their operational efficiency is really an emergence, and no 'process representation', no understanding of the collective goal, no grasp on the conditions for collective efficiency is needed at the individual agent level. The interaction of agent components produces an aggregate entity that is more flexible and adaptive than its component agents. Second, they develop and stabilise 'on the edge of chaos': they create order (invariants, rules, regularities, structures) against chaos, but they need residual disorder to survive. If no ant got lost while following the pheromone track to the prey, no new prey would be discovered soon enough. Too much order leads to crisis (e.g., epilepsy crises in the brain are due to a sudden synchronization of neuronal electrical activity). This leads to a vision of accidents as a resonance phenomenon (Hollnagel, 2004). So these systems/processes are not optimised for a given environment: they do not do the best possible job, but they do enough of the job. They are 'sufficient', to take René Amalberti's words elsewhere in this book (Chapter 16). They keep sub-optimal features as a trade-off against efficiency and the ability to cope with variation in their environment. Due to their auto-catalytic mechanisms, they have non-linear behaviour and go through cycles of stability and instability (like the stock market), but they are extremely stable (resilient) against aggressions, up to a certain threshold at which point they collapse. More precisely, their 'internal variables have been tuned to favour small losses in common events, at the expense of large losses when subject to rare or unexpected perturbations, even if the perturbations are infinitesimal' (Carlson & Doyle, 2002).

One of the variables that can be tuned is the topology of the system, for example the topology of communication structures. Barabási & Bonabeau (2003) have shown that scale-free communication networks are more resistant to random threat destruction than random networks, while they are more sensitive to targeted (malicious) attacks.

Scale-free (or scale-invariant) networks have a modular structure, rather than an evenly distributed one. Some of their nodes are highly connected, while others are poorly connected. Airlines 'hubs', and the World Wide Web (Figure 4.3) are example of such networks.

Recent work on network robustness has shown that scale-free architectures are highly resistant to accidental (random) failures, but very vulnerable to deliberate attacks. Indeed, highly connected nodes provide more alternative tracks between any two nodes in the network, but a deliberate attack against these nodes is obviously more disruptive.

Finally, Strongly Emergent (organizational) Resilience (SER) would refer to the (hypothetic) resilience features of an organization that cannot be explained by any combination of individual agent properties, even in principle. Although there is no evidence that such properties exist, recent developments in complexity science tend to prove that they are possible, provided a sufficient level of complexity is reached, which is undoubtedly the case for living beings (Davies, 2004). A candidate for this status may be the notion of culture, although it is disputable whether culture is a binding behavioural frame above individual behaviour, or merely a crystallised representation of actual behaviour. Another good candidate for this status would certainly be all the recursive self-reference phenomena, such as consciousness. Human social behaviour is increasingly self-represented through the development of mass media, hence partially recursive: anticipations modify what is anticipated (e.g., polls before a vote). This leads to circular causality and paradoxes (Dupuy & Teubner, 1990), and opens the door to some form of downward causation, or more accurately, to a kind of retroactive causation, so to speak: a representation of future can change the present. The certainty that a risk is totally eradicated will most probably trigger behaviours that will reintroduce the threat.

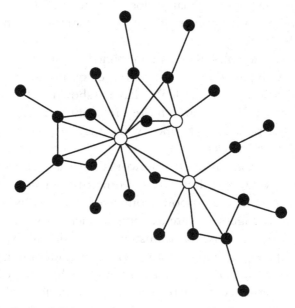

Figure 4.3: Scale-free networks

A key notion to understand the stabilization of circular causal dynamics is the notion of an attractor. An attractor is a fixed point in the space of possible states, for example a state of public opinion that is stable when opinion is informed of its current state. A particularly interesting kind of attractor has been known as the 'self-fulfilling prophecy', i.e., a representation of future that triggers, in the present time, actions and reactions which make that future happen. From a risk management perspective, the key question is how to keep concern for risk alive when things look safe. What we need to do is to introduce 'heuristics of fear' in order to 'build a vision of the future such that it triggers in the present time a behaviour preventing that vision from becoming real' (Dupuy, 2002). In other words, a safety manager's job is to handle irony: the core of a good safety culture is a self-defeating prophecy, and a whistle blower's ultimate achievement is to be wrong.

Conclusion

We have seen that a hierarchy of complexity can be described, each level corresponding to a specific notion of emergence. Human societies and organizations exhibit an overwhelming degree of complexity, and obviously cover the whole hierarchy. As an emerging property of these complex systems, resilience should also be considered through a similar taxonomy. Hence we have touched on differentiating nominal, weak and strong organizational resilience and discussed what these notions would cover. Standing back a little bit, it seems that most of the methodological efforts devoted to improve the robustness of organizations against safety threats, particularly within the industrial domain, have focused on nominal resilience. Compared to the actual complexity of the real world dynamics, the current approaches to safety management systems look terribly static and linear. It may be time for safety management thinkers and practitioners to look beyond the Heinrich (1931) domino model, and seek inspiration from complexity science and systems theory. This is, among other things, what this book is modestly but bravely trying to initiate.

Chapter 5

A Typology of Resilience Situations

Ron Westrum

'Resilience' involves many different matters covered by the same word. None the less I believe we can see some commonalities in the threats that resilience protects against and the modes in which it operates. The aim of this paper will be to sort out these threats and modes, which I believe provide different implications for developing resilient systems.

Resilience against What?

The first question to examine is the nature of the threats to the integrity of the system that require it to be resilient. There are basically three aspects to threats:

- The predictability of the threat. Predictability does not mean that we can predict when the event will take place, but only that it takes place fairly often.
- The threat's potential to disrupt the system.
- The origin of the threat (internal vs. external). If the origin is internal, we think of internal checks, safeguards or quality controls that would repair internal errors or pathogens and keep them from spreading. An external origin requires a response on the system's part. (Protection against threats are usually covered under the rubric of 'defenses', but this word does not distinguish between the two origins.)

With these basic aspects, let us see what we can do to develop a classification of situations. The situations described here are merely illustrative of the variety of scenarios likely to be met. Clearly, for practical affairs there might be reasons to develop a more exhaustive and complete set, thus a far more detailed tableau could be created.

Situation I. The Regular Threat

Regular threats are those that occur often enough for the system to develop a standard response. Trouble comes in one of a number of standard configurations, for which an algorithm of response can be formulated. Clearly, the least scary threat is the predictable one from inside the system, with low potential to disrupt the system as a whole. Medication errors, of which Patterson et al. (2004a) have given an excellent example, would fall into this category. This does not mean that such errors cannot be fatal. As the case study shows, they often have this potentiality. Nonetheless, they occur often, usually only implicate a single patient in the hospital, and potentially can be brought under control.

A similar example would be the following story by chemistry Professor Edgar F. Westrum, Jr, my father:

> About once a year, a chemistry student would go into one of the laboratories, start mixing chemicals at random, and eventually there would be a ka-whump! when one of the chemicals proved combustible when added to the others. People would dash to the lab to see what happened. The student would be taken to the hospital for repair, and the laboratory would be cleaned up.

At the next level we have the predictable external threat. This is more disturbing. In Israel bus bombings occur regularly enough to have allowed one of the hospitals to provide a standard response. Richard Cook, during the conference, described how the Kirkat Menachem hospital was able to shift resources between services to provide an effective response when a bus bombing occurred (Chapter 13). In countries where earthquakes are common, programmed responses, such as going outside, are also drummed into the local population.

Situation II. The Irregular Threat

The more challenging situation is the one-off event, for which it is virtually impossible to provide an algorithm, because there are so many similar low-probability but devastating events that might take place; and one cannot prepare for all of them. This kind of threat provides an understood, but still challenging problem. The Apollo 13 moon mission accident would be a good example here. When the event took place, it was entirely unexpected, though hardly seen as impossible. The spacecraft's spontaneous explosion and venting, though it did not destroy the whole structure, required the most drastic measures so that the astronauts could survive and return to earth. Fortunately, the Americans had a gifted team at NASA flight control, and so even though the mission was abandoned, the astronauts were brought safety back (Kranz, 2000).

While Israeli bus bombings typically fall into Situation I, other bombings, with far more casualties, belong in situation II. One such case was the suicide bombing of an American mess tent in Forward Operating Base Marez, in Iraq, on December 21, 2004. In 11 hours the 67th combat support hospital took in 91 casualties. There were 22 deaths. The halls were crowded, the parking lot overflowed with the dead and the dying, the conference room was turned into a blood donation center. Armored combat vehicles served as ambulances ferrying patients. Nine surgeries were carried out in a small operating theatre inside, while ten more had to be performed outside the operating room. At one point, a doctor began marking the time of death on the chests of the deceased with a felt-tip pen. They could barely get the dead into body bags fast enough. Finally, the hospital compound itself came under fire. The emergency was handled, but only just. This kind of emergency clearly tests the organization's ability to self-organise and respond effectively to crisis (Hauser, 2004).

Situation III. The Unexampled Event

Situation III is marked by events that are so awesome or so unexpected that they require more than the improvization of Situation II. They require a shift in mental framework. It may appear impossible that something like the event could happen. Whereas Situation II is basically a scale-up of Situation I, Situation III pushes the responders outside of their collective experience envelope. The 9/11 bombing of the World Trade Center is a prime example of such an event. The explosion at Chernobyl nuclear center is a similar example. An extreme example would be an invasion from outer space (see Dwyer, 2002).

Again unlike categories I and II, this level of threat cannot be anticipated neatly enough to permit construction of a response algorithm. Instead the basic qualities of the organization, its abilities to self-organize, monitor and formulate a series of responses, will determine whether it can react effectively. The response of both developed and undeveloped countries to the AIDS pandemic in the 1980s shows the great differences in coping styles of different countries (Institute of Medicine, 1995; Kennis & Marin, 1997).

Especially in the developing world, Situation III events can be devastating, because the level of resilience is so low. The third world both seems to take greater risks, and has fewer resources to respond when things go wrong. All this was horribly illustrated when the overloaded ferry Le Joola sank off Senegal on the night of 26 October 2002; 1863 people died. This was a larger death toll than the sinking of the Titanic, whose loss of life was 1500+. The ferry, which rolled over in a storm, had been loaded to triple its design capacity, the result of several unsafe bureaucratic practices, including a suicidal inability to limit entry to the ship. Fishermen saved 64 people after the accident, but there was no further rescue. Many of those who drowned were the victims of the incredibly slow response to the disaster. The saving of more survivors was imperilled by a corrupt bureaucracy. The radio operators on shore were not at their stations overnight, so the sinking was not noticed until the next day, when it was too late to rescue most of those who might have been saved by an early response. And even then help was further delayed (Fields, 2002). The approach of many third world countries to safety could well be described by the phrase 'risk-stacking,' since several types of vulnerability are taken on at once. Similarly, after the Bhopal fertilizer plant explosion in India, the government's response was so ineffective that many more people died than would otherwise have been the case (Shrivastava, 1987).

Time: Foresight, Coping, and Recovery

A second issue is: Where does the event lie on the organization's time horizon? Protecting the organization from trouble can occur proactively, concurrently, or as a response to something that has already happened. These are all part of resilience, but they are not the same. Resilience thus has three major meanings.

- Resilience is the ability to prevent something bad from happening,
- Or the ability to prevent something bad from becoming worse,
- Or the ability to recover from something bad once it has happened.

It is not clear at this point that possession of one of these capabilities automatically implies the others, or even that they form a Guttman scale, but either of these might be possible.

Foresee and Avoid

The ability to anticipate when and how calamity might strike has been called 'requisite imagination' (Adamski & Westrum, 2003). This ability, often important for the designer of technology, is also important for the operating organization. One might want to call this quality 'the wily organization.' The kind of critical thinking involved here has been studied in part by Irving Janis, famous for his analysis of the opposite trait, groupthink (Janis, 1982).

For the sake of analysis, let us examine two types of foresight. The first comes from properly learning the lessons of experience. In this case predictable threats are handled by programming the organization to remember the lessons learned. A recent study of Veterans' Administration physicians showed that they frequently forgot to use known effective methods of dealing with cardiac and pneumonia problems. The results were devastating; there have been thousands of unnecessary deaths. After many of the simple methods forgotten were reinstated, hospitals saw their relevant mortality rates drop by as much as 40% (Kolata, 2004).

The second type of foresight is that associated with the processing of 'faint signals.' These can include symptomatic events, suspected trends, gut feelings, and intelligent speculation. In another paper, I have suggested that 'latent pathogens' are likely to be removed in organizational cultures with high alignment, awareness, and empowerment. These boost the organization's capability to detect, compile and integrate diverse information. This capability helps in detecting 'hidden events' but also encourages proactive response to dangers that have not yet materialized.

Another striking example is the surreptitious repair of the Citicorp Building in New York City. Its architectural consultant, William LeMessurier, suspected and found that the structure had been built with changes to his original plan, using rivets instead of welds. A student's question stimulated LeMessurier to check his building's resistance to 'quartering winds,' those that attack a building at the corners instead of the sides. When it was clear that the changes were fateful, the architect, the contractor, and the police conspired to fix the building without public knowledge, in the end creating a sound structure without inciting a public panic or a serious financial loss for Citicorp (LeMessurier, 2005).

One of the key decisions affecting the Battle of Britain was taken by a private individual, Sidney Camm, head of the Hawker firm of aircraft builders. Even before the war with the Axis had begun, Camm realized that Britain would need fighters in the event of a German air campaign against England. He expected soon to get an order to build more Hurricane fighters. But rather than wait until the government ordered them, Camm commenced building them without a contract, the first production being in October 1937. In the Battle of Britain every single aircraft proved indispensable, and Hurricanes were the majority of those Britain used. Camm was knighted for this and other services to England (Deighton, 1996; org. 1978).

By contrast, organizations that indulge in groupthink and a 'fortress mentality' often resist getting signals about real or potential anomalies. This may take place in the context of faulty group discussion, as Irving Janis has forcefully argued (Janis, 1982), or through the broader organizational processes involving suppression, encapsulation, and public relations that I have sketched in several papers (e.g., Westrum & Adamski, 1999, pp. 97-98). During the time when the space shuttle Columbia was flying on its last mission, structural engineers at the Johnson Space Center tried to convince the Mission Management Team that they could not assure the shuttle's airworthiness. They wanted to get the Air Force to seek photographs in space that would allow them to make a sound decision. In spite of several attempts to get the photographs (which the Air Force was willing to provide), the Mission Management Team and other parties slighted the engineers' concerns and refused to seek further images. Interestingly enough, the reason for this refusal was often that 'even if we knew that there is damage, there is nothing we can do.' This assertion may have been justified, but one would certainly have liked to see more effort to save the crew (Cabbage & Harwood, 2004). One wonders what might have happened if a NASA flight director such as Gene Kranz had been in charge of this process.

It is tempting to contrast the striking individual actions of LeMessurier and Camm with the unsatisfactory group processes of the Columbia teams. However, the key point is the climate of operation, no matter how large the group is. There are individuals who personally can serve as major bottlenecks to decision processes and groups who become the agents of rapid or thoughtful action. It all depends on leadership, which shapes the climate and thus sets the priorities.

Coping with Ongoing Trouble

Coping with ongoing trouble immediately raises the questions of defences and capabilities. A tank or a battleship is resilient because it has armor. A football team is resilient because its players are tough and its moves are well coordinated. But often organizations are resilient because they can respond quickly or even redesign themselves in the midst of trouble. They may use 'slack resources' or other devices that help them cope with the struggle. The organization's flexibility is often a key factor in organizing to fight the problem. They are thus 'adaptive' rather than 'tough.' This is true, for instance, of learning during conflict or protracted crisis. Many examples of such learning are apparent from the history of warfare. There are many aspects to such learning. One aspect is learning from experience.

Such learning is the inverse of preconception. In the American Civil War, union cavalry with repeating rifles typically won the battles in which they were engaged. In spite of this experience, the repeating rifles were confiscated from cavalry units after the war and the cavalry were issued single-shot weapons. This was thought necessary to save ammunition. It did not serve American soldiers very well (Hallahan, 1994, pp. 198-199). Preconception can survive many severe lessons.

In World War II, British bureaucracy often got in the way of carrying on the war. This apparently was the case with preconceptions about the use of cannon. Gordon Welchman, in a book well worth reading, comments on British inability to use weapons in World War II for purposes other than their original one.

> In weaponry, too, the British thought at that time suffered from rigidity and departmentalization. The German 88-mm gun, designed as an anti-aircraft weapon, was so dangerous that four of them could stop an armored brigade. The British had a magnificent 3.7 inch anti-aircraft gun of even greater penetrative power, but it was not used as an antitank weapon in any of the desert battles. It was intended to shoot at aircraft. The army had been supplied with the two-pounder to shoot at tanks. And that was that!
>
> This is an example of the slow response of the British military authorities to new ideas. The Germans had used the 88-mm anti-aircraft gun as an extremely effective weapon against the excellent British Matilda tanks during the battle of France, as I well remember from German messages decoded by Hut 6 at the time. (Welchman, 1982, p. 229)

Another instance of inflexibility in World War II involved aircraft parts. During the Battle of Britain the supply of aircraft lagged disastrously. Some parts factories were slow in providing the Royal Air Force with new parts; other facilities were slow in recycling old ones. Lord Beaverbrook, as head of aircraft production, was assigned by Winston Churchill to address the problem. Beaverbrook's aggressive approach operated on many fronts. He rounded up used engines, staged illegal raids on parts factories, chloroformed guards, and got the necessary parts. He corralled parts into special depots and guarded them with a 'private army' that used special vehicles called 'Beaverettes' and 'Beaverbugs' to protect them. When Churchill received complaints about Beaverbrook's ruthless approach, he said he could do nothing. Air power was crucial to England, whose survival hung in the balance (Young, 1966, p. 143).

The ability to monitor what is happening in the organization is crucial to success in coping, whether this monitoring is coming from above or with the immediate group involved. The Report on 9/11 showed that coping with the attacks on the Pentagon and World Trade Center was often defective because of poor organizational discipline, previous unresolved police/fire department conflicts, and so forth (National Commission on Terrorist Attacks, 2004). For instance, New York's inability to create a unified incident command system, to resolve conflicts between police and fire department radio systems, and to insure discipline in similar emergencies meant a large number of unnecessary deaths. Fire-fighters who heard the distress calls often went to the scene, without integrating themselves into the command structure. This led to on-the-scene overload and confusion, and an inability of the supervising officials to determine how many fire-fighters were still in the structure. Similarly, after the failure of the south tower, the police department failed to inform the fire fighters inside the north tower that they knew that structure was also about to collapse (Baker, 2002).

Another aspect of coping is rooting out underlying problems that exist in the system. James Reason has called these problems 'latent pathogens' (Reason, 1990). While the origins of these latent pathogens are well known, less is known about the processes that lead to their removal. In a paper by the author, I have speculated about the forces that lead to pathogens being removed (Westrum, 2003). I have suggested that organizations that create alignment, awareness, and empowerment among the work force are likely to be better at removing latent pathogens, and are therefore likely to have a lighter 'pathogen load.' Organizations with fewer underlying problems will be better in coping because their defences are less likely to be compromised.

It is also worth noting that medical mix-ups such as the drug error described by Patterson et al. (2004a) would probably be improved by having a 'culture of conscious inquiry,' where all involved recognize the potentialities of a dangerous situation, and compensate for the dangers by added vigilance and additional checks (Westrum, 1991).

Repairing after Catastrophe

Resilience often seems to mean the ability to put things back together once they have fallen apart. The ability to rebound after receiving a surprise attack in war is a good measure of such resilience. When the Yom Kippur War started on the night of 6 October 1973, the Israelis had expected attack, but were still unprepared for the ferocity of the joint Syrian and Egyptian assault that took place. The Israeli Air Force, used to unopposed dominance of the skies, suddenly found itself threatened by hundreds of Arab surface-to-air missiles, large and small. The war would reveal that the Israel Air Force, often viewed as the best in the world, had three problems that had not been addressed: 1) it was poor at communicating with the ground forces, 2) it had inadequate battlefield intelligence and 3) it had not properly addressed the issue of ground support. Ehud Yonay comments that 'So elemental were these crises that, to pull its share in the war, the IAF not only had to switch priorities but literally reinvent itself in mid-fighting to perform missions it had ignored if not resisted for years' (Yonay, 1993, p. 345). Fortunately for the Israelis, they were able to do just that. Initiatives for ground support that had stalled were suddenly put into place, the Air Force was able to successfully coordinate its actions with the Israeli ground forces. Israel was able to repel the attacks against it. The ability to turn crisis into victory is seldom so clearly shown.

Recovery after natural disasters is often a function of the scale of the damage and the frequency of the type of disaster. Florida regularly experiences hurricanes, so there are routines used for coping. On the other hand, the Tsunami that recently engaged South East Asia was on such a colossal scale that even assistance from other countries came slowly and unevenly.

Recovery is easier for a society or an organization if the decision-making centers do not themselves come under attack. The 9/11 attack, on the World Trade Center, on the other hand, removed a local decision-making center in the United States. The attack would have done more damage if the attacks on the Pentagon and the White House had been more successful.

Conclusion

I have found that putting together this typology has helped me see some of the differences in the variety of situations and capabilities that we lump together under the label 'resilience.' No question that resilience is important. But what is it? Resilience is a family of related ideas, not a single thing. The various situations that we have sketched offer different levels of challenge, and may well be met by different organizational mechanisms. A resilient organization under Situation I will not necessarily be resilient under Situation III. Similarly, because an organization is good at recovery, this does not mean that the organization is good at foresight. These capabilities all deserve systematic study. Presumably, previous disaster studies have given coping and recovery much study. I regret that time did not permit me to consult this literature.

Organizational resilience obviously is supported by internal processes. I have been forced here to concentrate on the organization's behaviour and its outcomes. But many of the other papers in this volume (e.g., Chapter 14) look into the internal processes that allow resilience. There is little doubt that we need to know these processes better. In the past, the safety field as I know it has concentrated on weaknesses. Now perhaps we need to concentrate on strengths.

Acknowledgement

Many of the ideas for this chapter were suggested by other participants in the conference, and especially by John Wreathall.

Resilient Systems

Yushi Fujita

Engineers endow artifacts with abilities to cope with expected anomalies. The abilities may make the system robust. They are, however, a designed feature, which by definition cannot make the system 'resilient.' Humans at the front end (e.g., operators, maintenance people) are inherently adaptive and proactive; that allows them to accomplish better performances and, sometimes, even allows them to exhibit astonishing abilities in unexpected anomalies. However, this admirable human characteristic is a double-edged sword. Normally it works well, but sometimes it may lead to a disastrous end. Hence, a system relying on such human characteristics in an uncontrolled manner should not be called 'resilient.' A system should only be called 'resilient' when it is tuned in such a way that it can utilize its potential abilities, whether engineered features or acquired adaptive abilities, to the utmost extent and in a controlled manner, both in expected and unexpected situations.

Chapter 6

Incidents – Markers of Resilience or Brittleness?

David D. Woods
Richard I. Cook

Incidents are Ambiguous

The adaptive capacity of any system is usually assessed by observing how it responds to disruptions or challenges. Adaptive capacity has limits or boundary conditions, and disruptions provide information about where those boundaries lie and how the system behaves when events push it near or over those boundaries. Resilience in particular is concerned with understanding how well the system adapts and to what range or sources of variation. This allows one to detect undesirable drops in adaptive capacity and to intervene to increase aspects of adaptive capacity.

Thus, monitoring or measuring the adaptiveness and resilience of a system quickly leads to a basic ambiguity. Any given incident includes the system adapting to attempt to handle the disrupting event or variation on textbook situations. In an episode we observe the system stretch nearly to failure or a fracture point. Hence, are cases that fall short of breakdown success stories or anticipations of future failure stories? And if the disruption pushes the system to a fracture point, do the negative consequences always indicate a brittle system, since all finite systems can be pushed eventually to a breaking point?

Consider the cases in this book that provide images of resilience. Cook & Nemeth (Chapter 13, example 2) analyzes how a system under tremendous challenges has adapted to achieve high levels of performance. The medical response to attacks via bus bombings in

Israel quickly became highly adapted to providing care, to identifying victims, and to providing information and counselling to families despite many kinds of uncertainty and difficulties. Yet in that analysis one also sees potential limiting conditions should circumstances change.

Patterson et al. (2004b) analyze a case where the normal cross-checks broke down (one class of mechanisms for achieving resilience). As a result, a miscommunication and resulting erroneous treatment plan went forward with real, though not fatal, consequences for the patient involved. The medication misadministration in this case represents a breakdown in resilience of the system, yet in the analysis one also sees how the system did have mechanisms for reducing errors in treatment plans and cross-checks for detecting problems before they affected any patient (based on particular individuals who always handled the chemotherapy cases and thus had the knowledge and position in the network to detect possible problems and challenge physician plans effectively). In addition, the analysis of the case in terms of resilience factors revealed many aspects of successful and unsuccessful cross checks that could play out in other situations. By understanding the processes of cross-checking as a contributor to system resilience (even though they broke down in this episode), new system design concepts and new criteria for assessing performance are generated.

Cook & O'Connor (2005) provide a case that falls between the 'success' story of the Israeli medical response to bus bombings and the 'failure' story of the medication misadministration and the missed opportunities to catch that error. Cook & O'Connor describe the 'MAR knockout' (medication administration record) case where the pharmacy computer system broke down in a way that provided inaccurate medication plans hospital wide. The nurses did detect that medication plans were inaccurate, and that this was not an isolated problem but rather hospital wide. While no one knew why the medication administration records were wrong, the situation required a quick response to limit potential misadministrations to patients. Pharmacy, nurses, and physicians improvised a manual system by finding the last reliable medication records, updating these manually, and taking advantage of the presence of fax machines on wards. Through their efforts over 24 hours, no medication misadministrations occurred while the software trouble was diagnosed and restored (a failed reload from backup). The response to the computer infrastructure breakdown revealed that 'the resilience of this system resides in its people rather than in its technology. It was people who detected the failure, responded, planned a response, devised the workarounds needed to keep the rest of the system working, and restored the pharmacy system. There were a large cadre of experienced operations people ... available to assist in the recovery. ... What kept the MAR knockout case from being a catastrophe was the ability of workers to recover the system from the brink of disaster' (Cook & O'Conner, 2005).

While no one was injured, the authors see the case as an indicator of the increasing brittleness of the medication administration system. Computer infrastructure is becoming ever more vital to the ability to deliver care. Economic pressure is reducing the buffers, which operated in the incident response, as 'inefficiencies.' The resource of knowledgeable staff is tightening as managers misunderstand and undervalue the human contribution to successful function. Alternative modes of managing medications are becoming less viable as more equipment disappears (the goal of the 'paperless' hospital); and as the experience base for using alternative systems for managing medication orders is eroding.

Are cases of successful response to disruptions, stories of resilience that indicate future success if disruptions occur? Are stories of brittleness anticipations of future failures? Do the adaptations evident in any case illustrate resilience – since there was successful adaptation to handle the disrupting event, or do these cases reveal brittleness since the episode revealed how close the system came to falling off the edge? And the reverse questions can be posed as well for every case of failure: do breakdowns indict the adaptiveness of the system given that there are always finite resources, irreducible uncertainty, and basic dilemmas behind every situation?

These questions reveal that the value of incidents is in how they mark boundary conditions on the mechanisms/model of adaptiveness built into the system's design. Incidents simultaneously show how the system in question can stretch given disruptions and the limits on that capacity to handle or buffer these challenges.

Assessing resilience requires models of classes of adaptive behavior. These models need to capture the processes that contribute to adaptation when surprising disruptions occur. The descriptions of processes are useful when they specify the limits on these processes relative to challenges posed by external events. This allows resilience managers to monitor the boundary conditions on the current system's adaptive capacity and to target specific points where investments are valuable to preserve or expand that adaptive capacity given the vectors of change being experienced by that field of practice.

Patterson et al. (2004b) provide the start of one class of adaptive behavior – cross-check processes. Here we provide another class of adaptive behavior – the 'decompensation' event pattern.

'Decompensation:' A Pattern in Adaptive Response

One part of assessing a system's resilience is whether that system knows if it is operating near boundary conditions. Assessing the margin is not a simple static state (the distance of an operating point to a definitive boundary), but a more complex assessment of adaptive responses to different kinds of disturbances. Incidents are valuable because they provide information about what stretches the system and how well the system can stretch.

These complexities are illustrated by one kind of pattern in adaptive response called 'decompensation.' Cases of 'decompensation' constitute a kind of incident and have been analyzed in highly automated systems such as aircraft or cardiovascular physiology (Cook et al., 1991; Woods, 1994; Sarter et al., 1997). The basic decompensation pattern evolves across two phases. In the first phase, automated loops compensate for a growing disturbance; the successful compensation partially masks the presence and development of the underlying disturbance. The second phase of a decompensation event occurs because the automated response cannot compensate for the disturbance indefinitely. After the response mechanism's capacity is exhausted, the controlled parameter suddenly collapses (the de-compensation event that leads to the name).

The question is: does the supervisory controller of such systems detect the developing problem during the first phase of the event pattern or do they miss the signs that the lower order or base controller (automated loops in the typical system analysis) is working harder and harder to compensate and getting nearer to its capacity limit as the external challenge persists or grows?

In these situations, the critical information for the outside monitor is not the symptoms per se but the force with which they must be resisted relative to the capabilities of the base control systems. For example, when a human is acting as the base control system, as an effective team member this operator would communicate to others the fact that they need to exert unusual control effort (Norman, 1990). Such information provides a diagnostic cue for the team and is a signal that additional resources need to be injected to keep the process under control. If there is no information about how hard the base control system is working to maintain control in the face of disturbances, it is quite difficult to recognize the seriousness of the situation during the phase 1 portion and to respond early enough to avoid the decompensation collapse that marks phase 2 of the event pattern.

While this pattern has been noted in supervisory control of automated processes, it also can be used as an analogy or test case for the challenges of monitoring the resilience of organizations. In the standard decompensation pattern, the base system adapts to handle disruptions or faults. To determine if this adaptive behavior is a signal of successful control or a sign of incipient failure requires an assessment of the control capability of the base system in the face of various kinds and sizes of disturbances.

Consider the challenge for a person monitoring these kinds of situations. When disturbances occur, the presence of adaptive capacity produces counter-influences, which makes control appear adequate or allows only slight signs of trouble over a period of time. Eventually, if no changes are made or assistance injected, the capacity to compensate becomes exhausted and control collapses in the form of a major incident or accident. The apparent success during the first phase of the event can mask or hide how adaptive mechanisms are being stretched to work harder to compensate and how buffers are being exhausted. This makes it difficult for monitors to understand what is occurring, especially as the first phase may play out over relatively longer time periods, and leads to great surprise when the second phase occurs.

While detecting the slide toward decompensation is in principle difficult, decompensation incidents reveal how human supervisory controllers of automated systems normally develop the expertise needed to anticipate the approach to boundaries of adaptive capacity and intervene to provide additional capacity to cope with the disturbances underway. For example, in intensive care units and in anaesthesiology (Cook et al., 1991), physicians are able to make judgements about the adaptive capacity of a patient's physiological systems. Precariously balanced heart patients have little capacity for handling stressors or disrupting events – their control mechanisms are 'run out.' Anesthesiologists and intensivists try to avoid physiological challenges in such cases and try to provide extra control capacity from outside mechanisms with drips of various cardiovascular medications. Note that adaptive behavior still occurs whether the patient is a young adult with a single problem (possesses a large adaptive reservoir) or a precariously balanced elderly patient with several longstanding cardiovascular related problems (possessing very limited control reserves). But there is something about the kinds of adaptations going on or not going on that allows the experienced physician to recognize the difference in adaptive capacities to project the cardiovascular system's ability to handle the next challenge (i.e., more than simple demographics and previous medical history).

There is information available to support the above kinds of assessments. The information consists of relationships between the state of the process being controlled, the disturbances the process has been or could be subjected to, and the kinds of responses that have been made to recent challenges. These relationships indicate whether control loops are having trouble handling situations (working very hard to maintain control), working at the extreme of their range, or moving towards the extreme part of their capabilities. Picking up on these relationships indicates to the observer whether or not there is limited remaining capacity to compensate even though no direct indicators of control failures will have occurred yet.

It is also clear that expert judgements about adaptive capacities come from tangible experience with the systems responses to variations and disruptions. Seeing how the system responds to small perturbations provides the feedback to the physician about the capacity to handle other and larger disruptions or challenges (a feel for the dynamics of disturbance and response often captured in terms like a 'sluggish' response). Such judgements require baseline information about normal adaptive capabilities given typical disturbances. In monitoring patient physiology, anesthesiologists may sometimes even inject small challenges to observe the dynamic responses of the cardiovascular system as a means to generate information about adaptive capacity.

Our conjecture is that, inspired directly or indirectly by these very detailed situations of judging adaptive capacity in supervisory control, we can create mechanisms to monitor the adaptive capacity of organizations and anticipate when its adaptive capacity is precarious.

For example, in safety management, change and production pressure are disturbances that erode or place new demands on adaptive capacity. During the first phase of a decompensation pattern, it is difficult for normal monitoring mechanisms to see or interpret the subtle signs of reduced adaptive capacity. Eventually, failures occur if adaptive capacity continues to erode and as new challenges combine. These failures represent a sudden collapse in adaptive capacity as demands continued to stretch the system. While a failure may be seen as a message about specific vulnerabilities and proximal factors, they are also signs of the general decompensation signature as a pattern of change in the adaptive capacity of the organization, cf. Doyle's 'HOT' analysis of systems with 'highly optimized tolerance' (Carlson & Doyle, 2000).

Returning to the opening question of this chapter, cases are not valuable based on whether adaptation was successful or unsuccessful. Cases are valuable because they reveal patterns in how adaptive capacity is used to meet challenges (Cook et al., 2000). Such cases need to be analyzed to determine when challenges are falling outside or stressing the competence envelope, how the system engages pools of adaptive capacity to meet these challenges, and where the adaptive capacity is being used up to handle these unanticipated perturbations. The decompensation kind of incident reminds us that the basic unit of analysis for assessing resilience is observations of a system's response to varying kinds of disturbances.

Measures of brittleness and resilience will emerge when we abstract general patterns from specific cases of challenge and response. Resilience engineering needs to search out incidents because they reveal boundary conditions and how the system behaves near these boundaries. These cases will provide evidence of otherwise hidden sources for adaptiveness and also illustrate the limits on a system's adaptive capabilities. General patterns in adaptive system behavior (as illustrated by the decompensation event signature and in other patterns taken from the cases of resilience/brittleness captured in this book) will help measure an organization's resilience and target interventions to enhance an organization's resilience.

Acknowledgements

This work was supported in part by grant NNA04CK45A from NASA Ames Research Center to develop resilience engineering concepts for managing organizational risk.

Chapter 7

Resilience Engineering: Chronicling the Emergence of Confused Consensus

Sidney Dekker

My contribution to this book on resilience engineering consists of a chapter that attempts to capture the emergence and exchange of ideas during the symposium (held in Söderköping, Sweden, during October 2004) – ideas that together seem to move towards a type of confused consensus on resilience engineering (and sometimes towards a common understand of what it is not). From the presentations, arguments, claims, hypotheses, questions and suggestions, I have extracted themes that seemed to animate many of the ideas that generated discussions, and discussions that generated ideas. The themes, as organised below, could represent at least some kind of logical flow; an order in which one idea gives rise to the next, as the constraints of the previous one on progress for resilience engineering become clear. I first present the themes in summary, and then get into them in more detail:

- We have to get smarter at predicting the next accident. The recombinant predictive logic that drives accident prediction today is insensitive to the pressures of normal work by normal people in normal organizations. We may not even need accident models, as the focus should be more on models of normal work. It is normal work (and the slow, incremental drift into safety margins and eventually across boundaries that is a side effect of normal work in resource-constrained, adaptive organizations) that seems to be an important engine behind complex system accidents.
- Detecting drift into failure that happens to seemingly safe systems, *before* breakdowns occur is a major role for resilience engineering.

While it is critical to capture the relational dynamics and longer-term socio-organizational trends behind system failure, 'drift' as such is ill-defined and not well-modelled. One important ingredient in drift is the sacrificing decision, where goals of safety and production/efficiency are weighed against one another, and managers make locally rational decisions based on criteria from inside their situated context. But establishing a connection between individual micro-level managerial trade-offs and macro-level drift is challenging both practically and theoretically. In addition, it is probably not easy to influence sacrificing decisions so that they do not put a system on the road towards failure. Should we really micro-adjust managers' day-to-day cognitive calculus of reward?

• Detecting drift is difficult and makes many assumptions about our ability to distinguish longitudinal trends, establish the existence of safety boundaries, and warn people of their movement towards them. Instead, a critical component in estimating an organization's resilience could be a momentary charting of the distance between operations as they really go on, and operations as they are imagined in the minds of managers or rulemakers. This distance tells us something about the models of risk currently applied, and how well (or badly) calibrated organizational decision makers are. This, however, requires comparisons and perhaps a type of quantification (even if conceptual) that may be challenging to attain. Updating management about the real nature of work (and their deficient perception of it) also requires messengers to go beyond 'reassuring' signals to management, which in turn demands a particular climate that allows the boss to hear such news.

• Looking for additional markers of resilience, what was explored in the symposium? If charting the distance between operations as imagined and as they really occur is too difficult, then an even broader indicator of resilience could be the extent to which the organization succeeds in keeping discussions of risk alive even when everything looks safe. The requirement to keep discussions of risk alive invites us to think about the role and nature of a safety organization in novel ways. Another powerful indicator is how the organization responds to failure: can the organization rebound even when exposed to enormous pressure? This closes the circle as it brings us back to theme 1, whether resilience is about effectively predicting (and then preventing) the next accident. In terms of the cognitive challenge it represents, preventing the next accident could conceptually be close to managing the aftermath of one: doing both effectively involves rebounding from previous understandings of the system and interpretations of the risk it is exposed to, being forced to see the gap between work as imagined and work as done, updating beliefs about safety and brittleness, and recalibrating models of risk.

Resilience Engineering and Getting Smarter at Predicting the Next Accident

The need for resilience engineering arises in part out of the inadequacy of current models for understanding, and methods for predicting, safety breakdowns in complex systems. In very safe systems, incident reporting no longer works well, as it makes a number of questionable assumptions that no longer seem to be valid.

The first assumption of incident reporting is that we possess enough computational capacity to perform meaningful recombinant logic on the database of incidents, or that we have other working ways of extracting intelligence from our incident data. This assumption is already being outpaced by very successful incident reporting systems (e.g., NASA's Aviation Safety Reporting System) which indeed could be said to be a victim of their own success. The number of reports received can be overwhelming, which makes meaningful extraction of risk factors and prediction of how they will recombine extremely difficult. In fact, a main criterion of success of incident reporting systems across industries (the number of reports received) could do with a reconsideration: celebrating a large count of negatives (instances of system problems, errors, or failures) misses the point since the real purpose of incident reporting is not the gathering of data but the analysis, anticipation and reduction of risk.

The second assumption is that incident reports actually contain the ingredients of the next accident. This may not always be true, as accidents in very safe systems seem to emerge from (what looks to everybody like) normal people doing normal work in normal organizations. In the time leading up to an accident there may be few reportworthy failures or noteworthy organizational deficiencies that would end up as recombinable factors in an incident reporting system. Even if we had the computational capacity to predict how factors may recombine to produce an accident, this is going to be of no help if the relevant 'factors' are not in the incident database in the first place (as they were never judged to be 'abnormal' enough to warrant reporting). The focus on human errors and failure events may mean that incident reporting systems are not as useful when it comes to predicting accidents in safe systems. This is because in safe systems, it is not human errors or failure events that lead to accidents. Normal work does.

The third assumption is that the decomposition principles used in incident reporting systems and the failure models that underlie them have something to do with the way in which factors actually recombine to create an accident. Most fundamentally, the metaphors that inspire and justify incident reporting (e.g., the 'iceberg', the 'Swiss cheese'), assume that there is a linear progress of breakdowns (a domino effect) that either gets stopped somewhere (constituting an incident) or not (creating an accident). Neither the presumed substantive similarity between incident and accident, nor the linear, chain-like breakdown scenarios enjoy much empirical support in safe systems. Instead, safety and risk in safe systems are emergent properties that arise from a much more complex interaction of all factors that constitute normal work.

As a result, linear extensions of current safety efforts (incident reporting, safety and quality management, proficiency checking, standardization and proceduralization, more rules and regulations) seem of little use in breaking the asymptote, even if they could be necessary to sustain the high safety levels already attained.

Do We Need Accident Models at All?

The problem with the third assumption calls another fundamental idea into question: what kinds of accident models are useful for resilience engineering? People use models of accidents the whole time. Most accident models, especially naïve or folk ones, appeal to a kind of exclusivity of factors and activities (or how they interact) that lead up to an accident. The idea is that organizational work (or the organizational structure that exists) preceding an accident is somehow fundamentally or at least identifiably different from that which is not followed by an accident. Thus, accidents require their own set of models if people want to gain predictive leverage. Accident models are about understanding accidents, and about presumably predicting them before they happen. But if success and failure stem from the same source, then accident models are either useless, or as useless (or useful) as equally good models of success would be. The problem is that we do not have many good models of operational success, or even of normal work that does not lead to trouble. More than once, symposium participants remarked (in multifarious language) that we need better models of normal work, including better operationalised models of the effects of 'safety culture' (or lack thereof) on people's everyday decision-making.

Resilience engineering wants to differ from safety practices that have evolved differently in various industries. The nuclear industry, for example, relies to a large extent on barriers and defences in depth. Whatever goes on behind all the barriers and how risk may change or shift out there, the system is insulated from it. Aviation seems forever rooted in a 'fly-fix-fly' model, where (presumably or hopefully) fail-safe designs are introduced into operation. Subsequently, huge transnational, partially overlapping webs of failure monitoring (from event reporting to formal accident investigation) are relied upon largely 'bottom-up' to have the industry make adjustments where necessary. Aviation system safety thus seems to run relatively open-loop and represents a partially reactive control of risk. Military systems, having to deal with active adversaries, emphasise hazards, hazard analysis and hazard control, relying in large part on top-down approaches to making safer systems. In contrast with aviation, this represents a more proactive control of risk. But how do these systems control (or are even aware of) their own adaptive processes that enable them to update and adjust their management of risk?

There is a complex dynamic underneath the processes through which organizations decide how to control risk. One critical aspect is that doing things safely is ongoing. Consistent with the challenge to the use of accident models, doing things safely is perhaps not separable from doing things at all. Individuals as well as entire industries continually develop and refine strategies that are sensitive to known or likely pathways to failure, they continually apply these strategies as they go about their normal work, and they adapt these strategies as they see new evidence on risk come in. Leverage for change and improvement, then, lies in the interpretation of what constitutes such risk; leverage lies in helping individuals and systems adapt effectively to a constantly changing world, calibrating them to newly emerging threats, identifying obsolete models of risk, and helping them avoid overconfidence in the continued effectiveness of their strategies. This, as said above, could be central to resilience engineering: helping organizations not with better control of risk per se, but helping them better manage the processes by which they decide how to control such risk.

This obviously lies beyond the prediction of accidents. This lies beyond the mere controlling of risk, instead taking resilience engineering to a new level. It is about attaining better control of the control of risk. Rather than just helping organizations, or people, become sensitive to pathways to failure, resilience engineering is about helping them become sensitive about the models of risk that they apply to their search and control of pathways to failure.

Modelling the Drift into Failure

One way to think about the greatest risk to today's safe socio-technical systems is the drift into failure, and a potential contribution from resilience engineering could be to help organizations detect this drift. 'Drifting into failure' is a metaphor for the slow, incremental movement of systems operations toward (and eventually across) the boundaries of their safety envelope. Pressures of scarcity and competition typically fuel such movement and uncertain technology and incomplete knowledge about where the boundaries actually are, result in people not stopping the movement or even seeing it. Recognising that a system is drifting into failure is difficult because the entire protective structure (including suppliers, regulators, managerial hierarchies, etc.) seems to slide along with the operational core toward the boundary. Even if an operational system is 'borrowing more from safety' than it was previously or than it is elsewhere by operating with smaller failure margins, this may be considered 'normal', as the regulator approved it. Almost everybody inside the system does it, goes along, and agrees, implicitly or not, with what is defined as risky or safe. Also, the departures from previous practice are seldom quick or large or shocking (and thus difficult to detect): rather, there is a slow succession of tiny incremental deviations from what previously was the 'norm'. Each departure in itself is hardly noteworthy. In fact, such 'departures' are part and parcel of normal adaptive system behaviour, as organizations (and their regulators) continually realign themselves and their operations with current interpretations of the balance between profitability and risk (and *have* to do so in environments of resource scarcity and competition). As a result, large system accidents of the past few decades have revealed that what is considered 'normal', or acceptable risk is highly negotiable and subject to all kinds of local pressures and interpretations.

Do We Need (and Can We Attain) a Better Model of Drift?

The intuitive appeal of 'drift into failure' is not matched by specific, working models of it. In fact, drift into failure is an unsatisfactory explanation so far because it is not a model, just like the 'Swiss cheese' (or defences in depth) idea is not a model. It may be a good metaphor for how complex systems slowly move towards the edges of breakdown, but it lacks all kinds of underlying dynamic relationships and potentials for operationalization that could make it into a model.

One important ingredient for such a model would be the 'sacrificing decision'. To make risk a proactive part of management decision-making requires ways to know when to relax the pressure on throughput and efficiency goals, i.e., making a *sacrifice* decision, and resilience engineering could be working on ways to help organizations decide when to relax production pressure to reduce risk. Symposium participants agree on the need to better understand this judgement process in organizations, and how micro-level decisions are linked to macro-level 'drift'. The decision to value production over safety is often implicit, and any larger or broader side effects that echo through an organization go unrecognised. So far, the result is individuals and organizations acting much riskier than they would ever desire, or than they would have in retrospect. The problem with sacrificing decisions is that they are often very sound when set against local judgement criteria; given the time and budget pressures and short-term incentives that shape decision-making behaviour in those contexts. In other words, given the decision makers' knowledge, goals, focus of attention, and the nature of the data available to them at the time, it made sense. Another problem with the sacrifice judgement is related to hindsight. The retrospective view will indicate that the sacrifice or relaxation may have been unnecessary since 'nothing happened.' A substitute measure for the 'soundness' of a sacrificing decision, then, could be to assess how peers and superiors react to it. Indeed, for the lack of such *ersatz* criteria, aggregate organizational behaviour is likely to move closer to safety margins, the final evidence of which could be an accident.

Of course, it could be argued that 'sacrifice decisions' are not really special at all. In fact, all decisions are sacrifices of one kind or another, as the choosing of one option almost always excludes other options. All decisions, in that sense, could be seen as trade-offs, where issues of sensitivity and decision criterion enter into consideration (whether consciously or not).

The metaphor of drift, suggests that it is in these normal, day-to-day processes of organizational management and decision-making that we can find the seeds of organizational failure and success, and a role of resilience engineering could be to find leverage for making further progress on safety by better understanding and influencing these processes. One large operationalization problem lies in the reciprocal macro-micro connection: how do micro-level decisions and trade-offs not only represent and reproduce macro-structural pressures (of production, resource scarcity, competition) and the associated organizational priorities and preferences, but how are micro-level decisions related to eventual macro-level organizational drift? These links, from macro-structural forces down to micro-level decisions, and in turn from micro-level decisions up to macro-level organizational drift, are widely open for future investigation.

Yet even if we conduct such research, is a model of drift desirable? The complexity of the organizational, political, psychological and sociological phenomena involved would make it a serious challenge, but that may not even be the main problem. It would seem, as has been suggested before, that safety and risk in complex organizations are emergent, not resultant, properties: safety and risk cannot be predicted or modelled on the basis of constituent components and their interactions. So why try to build a model of drift as an emergent phenomenon, or why even apply modelling criteria to the way in which you want to capture it? One way to approach the way in which complex systems fail or succeed, then, is through a *simile* (like 'drift into failure') rather than through a model. A simile (just like the Swiss cheese) can guide, direct attention, take us to interesting places for research and progress on safety. It may not 'explain' as a traditional model would. But, again, this is the whole point of emergent processes – they call for different kinds of approaches (e.g., system dynamics) and different kinds of accident models.

Where are the Safety Boundaries that You can Drift Across?

Detecting how an organization is drifting into failure *before* breakdowns occur is difficult. One reason is that we lack a good, operationalised model of drift that could focus us on the tell-tale signs. And, as mentioned previously, a true model of drift may be out of reach altogether since it may be fundamentally immeasurable. Another issue is that of boundaries: drifting would not be risky if it were not for the existence of safety boundaries across which you actually *can* drift into breakdown. Other than shedding light on the adaptive life of organizations (a key aspect of resilience engineering), perhaps there is even no point in understanding drift if we cannot chart it relative to safety boundaries. But if so, what are these boundaries, and *where* are they?

The idea of safety boundaries appeals to intuitive sense: it can appear quite clear when boundaries are crossed (e.g., in an accident). Yet this is not the same as trying to plot them before an accident. There, the notion of hard, identifiable boundaries is very difficult to develop further in any meaningful way. With safety (and accidents) as emergent phenomena, deterministic efforts to nail down borders of safe practice seems rather hopeless. The number of variables involved, and their interaction, makes the idea of safety boundaries as probability patterns more appealing: probability patterns that vary in an indeterministic fashion with a huge complex of operational and organizational factors.

Going even further, it seems that safety boundaries, and where they lie, are a projection or an expression of our current models of risk. This idea moves away from earlier, implied realist interpretations of safety margins and boundaries, and denies their independent existence. They are not present themselves, but are reflections of how we think about safety and risk at that time and in that organization and operation. The notion of boundaries as probability patterns that arise out of our own projections and models of risk is as profoundly informative as it is unsettling. It does not mean that organizations cannot adapt in order to stay away from those boundaries, indeed, they can. People and organizations routinely act on their current beliefs of what makes them safe or unsafe. But it does mean that the work necessary to identify boundaries must shift from the engineering-inspired calculative to the ethnographically or sociologically interpretive. We want to learn something meaningful about insider interpretations, about people's changing (or fixed) beliefs and how they do or do not act on them.

Work as Imagined versus Work as Actually Done

One marker of resilience that comes out of converging lines of evidence, is the distance between operations as management imagines they go on and how they actually go on. A large distance indicates that organizational leadership may be ill-calibrated to the challenges and risks encountered in real operations.

Commercial aircraft line maintenance is emblematic: A job-perception gap exists where supervisors are convinced that safety and success result from mechanics following procedures – a sign-off means that applicable procedures were followed. But mechanics may encounter problems for which the right tools or parts are not at hand; the aircraft may be parked far away from base. Or there may be too little time: Aircraft with a considerable number of problems may have to be turned around for the next flight within half an hour. Mechanics, consequently, see success as the result of their evolved skills at adapting, inventing, compromising, and improvising in the face of local pressures and challenges on the line – a sign-off means the job was accomplished in spite of resource limitations, organizational dilemmas, and pressures. Those mechanics that are most adept are valued for their productive capacity even by higher organizational levels. Unacknowledged by those levels, though, are the vast informal work systems that develop so mechanics can get work done, advance their skills at improvising and satisficing, impart them to one another, and condense them in unofficial, self-made documentation (McDonald *et al.*, 2002). Seen from the outside, a defining characteristic of such informal work systems would be routine violations of procedures (which, in aviation, is commonly thought to be 'unsafe'). But from the inside, the same behaviour is a mark of expertise, fuelled by professional and interpeer pride. And of course, informal work systems emerge and thrive in the first place because procedures are inadequate to cope with local challenges and surprises, and because procedures' (and management's) conception of work collides with the scarcity, pressure and multiple goals of real work.

International disaster relief work similarly features widely diverging images of work. The formal understanding of relief work, which needs to be coordinated across various agencies and nations, includes an allegiance to distant supervisors (e.g., in the head office in some other country) and their higher-order goals. These include a sensitivity to the political concerns and bureaucratic accountability that form an inevitable part of cross-national relief work. Plans and organizational structures are often overspecified, and new recruits to disaster work are persuaded to adhere to procedure and protocol and defer to hierarchy. Lines of authority are clear and should be checked before acting in the field.

None of this works in practice. In the field, disaster relief workers show a surprising dissociation from distant supervisors and their global goals. Plans and organizational structures are fragile in the face of surprise and the inevitable *contretemps*. Actual work immediately drifts away from protocol and procedure, and workers defer to people with experience and resource access rather than to formal hierarchy. Improvisation occurs across organizational and political boundaries, sometimes in contravention to larger political constraints or sensitivities. Authority is diffuse and often ignored. Sometimes people check after they have acted (Suparamaniam & Dekker, 2003).

Generally, the consequence of a large distance appears to be greater organizational brittleness, rather than resilience. In aircraft maintenance, for example, the job perception gap means that incidents linked with maintenance issues do not generate meaningful pressure for change, but instead produce outrage over widespread violations and cause the system to reinvent or reassert itself according to the official understanding of how it works (rather than how it actually works). Weaknesses persist, and a fundamental misunderstanding of what makes the system work (and what might make it unsafe) is reinforced (i.e., follow the procedures and you will be safe). McDonald et al., 2002 famously called such continued dysfunctional reinvention of the system-as-it-ought-to-be 'cycles of stability'. Similarly, the distributed and politically pregnant nature of international disaster relief work seems to make the aggregate system immune to adapting on a global scale. The distance between the formal and the actual images of work can be taken as a sign of local worker intransigence, or of unusual adversity, but not as a compelling indication of the need to better adapt the entire system (including disaster relief worker training) to the conditions that shape and constrain the execution of real work. The resulting global brittleness is compensated for by local improvisation, initiative and creativity.

But this is not the whole story. A distance between how management understands operations to go on and how they really go on, can actually be a marker of resilience at the operational level. There are indications from commercial aircraft maintenance, for example, that management's relative lack of clues of about real conditions and fluctuating pressures inspires the locally inventive and effective use of tools and other resources vis-à-vis task demands. Similarly, practitioners ranging from medical workers to soldiers can 'hide' available resources from their leadership and keep them in stock in order to have some slack or reserve capacity to expand when demands suddenly escalate. 'Good' leadership always extracts maximum utilization from invested resources. 'Good' followership, then, is about hiding resources for times when leadership turns out to be not that 'good' after all. Indeed, leadership can be ignorant of, or insensitive to, the locally fluctuating pressures of real work on the sharp end. As a result, even safety investments can quickly get consumed for production purposes. In response, practitioners at the sharp end may secure a share of resources and hold it back for possible use in more pressurised times.

From an organizational perspective, such parochial and subversive investments in slack are hardly a sign of effective adjustment, and there is a need to better capture processes of cross-adaptation and their relationship to organizational resilience. One idea here is to see the management of resilience as a matter of balancing individual resilience (individual responses to operational challenges) and system resilience (the large-scale autocatalytic combination of individual behaviour). Again, the hope here is that this can produce a better understanding of the micro-macro connection – how individual trade-offs at the sharp end (themselves influenced by macro-structural forces) in turn can create global side effects on an emergent scale.

Measuring and Closing the Gap

How can we measure this gap between the system as designed or imagined and the system as actually operated? And, perhaps more importantly, what can we do about it? The distance between the real system and the system as imagined grows in part through local feedback loops that confirm managers into thinking that what they are doing is 'right'. Using their local judgement criteria, this can indeed be justified. The challenge then, is to make the gap visible and provide a basis for learning and adaptation where necessary. This implies the ability to contrast the present situation with, for example, a historical 'ideal'. The hierarchical control model proposed by Nancy Leveson is one example of an approach that can provide snapshots of the system-as-designed and the system as actually evolved. From one time to the next, it can show that changes have occurred in, and erosion has happened to, control loops between various processes and structures (Leveson, 2002). The gradual erosion of control, and the subsequent mismatch between the system-as-designed and the system as actually operated, can be seen as an important ingredient in the drift into failure.

One problem here is that operations 'as imagined' are not always as amenable to capture or analysis as the idealised control model of a system's starting position would be. Understanding the gap between the system-as-imagined and the system as actually operated requires investment not only in understanding how the system really works, but also how it is imagined to work. The latter can sometimes even be more difficult. Take the opinion of the quality manager of an airline flight school that suffered a string of fatal accidents, who 'was convinced that the school was a top training institution; he had no knowledge of any serious problems in the organization' (Raad voor de Transportveiligheid, 2003, p. 41). This is a position that can be shown, with hindsight, to contrast with a deeply troubled actual organization which faced enormous cost pressure, lacked qualified or well-suited people in leadership posts, ran operations with non-qualified instructors, featured an enormous distrust between management and personnel, and had a non-functioning reporting system (all of which was taken as 'normal' or acceptable by insiders, as well as the regulator, and none of which was taken as signals of potential danger). But the issue here is that the quality manager's idea about the system is not only a sharp contradistinction from the actual system, it is also very underspecified, general, vague. There is no way to really calibrate whether this manager has the 'requisite imagination' (Adamski & Westrum, 2003) to adapt his organization as necessary, although it is inviting to entertain the suspicion that he does not.

Ways forward on making both images of work explicit include their confrontation with each other. A part of resilience engineering could be to help leadership with what has been called 'broadening checks' – a confrontation with their beliefs about what makes the system work (and safe or unsafe) in part through a better visualization of what actually goes on inside. One approach to broadening checks suggested and developed by Andrew Hale is to do scenario-based auditing. Scenario-based auditing represents a form of guided simulation, which could be better suited for capturing eroded portions of an organization's control structure. In general, broadening checks could help managers achieve greater requisite imagination by confronting them with the way in which the system actually works, in contrast to their imagination.

Towards Broader Markers of Resilience

Looking for additional markers of resilience, what was explored in the symposium? Here I take up two points: whether safe organizations succeed in keeping discussion about risk alive, and how organizations respond to failure. Both of these could be telling us something important, and something potentially measurable, about the resilience or brittleness of the organization.

Keeping Discussions about Risk Alive Even When all Looks Safe

As identified previously, an important ingredient of engineering a resilient organization is constantly testing whether ideas about risk still match with reality; whether the model of operations (and what makes them safe or unsafe) is still up to date. As also said, however, it is not always easy to expose this gap between the system-as-imagined and the system as really operated. The models that managers and even operational people apply in their management of risk could be out of date without anybody (including outsiders, such as researchers or regulators) knowing it because they are often so implicit, so unspecified. In fact, industries are seldom busily entertained with explicating and recalibrating their models of risk. Rather, there seems to be hysteresis: old models of risk and safety (e.g., human errors are a major threat to otherwise safe systems) still characterise most approaches in aviation, including latter-day proposals to conduct 'line-oriented safety audits' (which, ironically, are designed explicitly to find out how real work actually takes place!).

If the distance between the model of risk (or the system as imagined) and actual risk (or the system as actually operated) is difficult to illuminate, are there other indicators that could help us judge the resilience of an organization? One way is to see whether activities associated with recalibrating models of safety and risk are going on at all. This typically involves stakeholders discussing risk even when everything looks safe. Indeed, if discussions about risk are going on even in the absence of obvious threats to safety, we could get some confidence that an organization is investing in an analysis, and possibly in a critique and subsequent update, of its models of risk. Finding out that such discussions are taking place, of course, is not the same as knowing whether their resulting updates are effective or result in a better calibrated organization. It is questionable whether it is sufficient just to keep a discussion about risk alive. We should also be thinking about ways to measure the extent to which these discussions have a meaningful effect on how the organization relates to operational challenges. We must remind ourselves that talking about risk is not the managing of it, and that it constitutes only a step toward actually engineering the resilience of an organization.

The role of a safety organization in fanning this discussion, by the way, is problematic. We could think that a safety department can take a natural lead in driving, guiding and inspiring such discussions. But a safety department has to reconcile competing demands for being involved as an insider (to know what is going on), while being independent as an outsider (to be able to step back and look at what really is going on and influence things where necessary). In addition, a strong safety department could send the message to the rest of an organization that safety is something that can and should be delegated to experts; to a dedicated group (and indeed, parts of it probably should be). But keeping discussions about risk alive is something that should involve every organizational decision maker, throughout organizational hierarchies.

Resilience and Responses to Failure

What if failures do occur? Are there criteria of resilience that can be extracted then? Indicators of how resilient an organization is consistently seem to include how the organization responds to failure. Intuitively, crises could be a sign of a lack of fundamental organizational resilience, independent of the effectiveness of the response. But effective management of crises may also represent an important marker of resilience: What is the ability of the organization to rebound even when exposed to enormous pressure? Preventing the next accident could conceptually be close to managing the aftermath of one and in this sense carry equally powerful implications for our ability to judge the resilience of an organization.

The comparison is this: suppose we want to predict an accident accurately. This, and responding to a failure effectively, both involve retreating or recovering from previous understandings of the system and interpretations of the risk it is exposed to. Stakeholders need to abandon old beliefs about the real challenges that their system faces and embrace new ones. Both are also about stakeholders being forced to acknowledge the gap between work as imagined and work as done. The rubble of an accident often yields powerful clues about the nature of the work as actually done and can put this in sharp contradistinction with official understandings or images of that work (unless the work as actually done retreats underground until the probe is over – which it does in some cases, see McDonald et al., 2002). Similarly, getting smarter at predicting the next accident (if we see that as part of resilience engineering) is in part about finding out about this gap, this distance between the various images of work (e.g., 'official' versus 'real'), and carefully managing it in the service of better grasping the actual nature of risk in operations.

Cognitively, then, predicting the next accident and mopping up the previous one demands the same types of revisionist activities and insights. Once again, throughout the symposium (and echoed in the themes of the present book) this has been identified as a critical ingredient of resilience: constantly testing whether ideas about risk still match reality; updating beliefs about safety and brittleness; and recalibrating models of risk. A main question is how to help organizations manage these processes of insight and revision effectively, and further work on resilience engineering is bound to address this question.

Part II: Cases and Processes

Chapter 8

Engineering Resilience into Safety-Critical Systems

Nancy Leveson
Nicolas Dulac
David Zipkin
Joel Cutcher-Gershenfeld
John Carroll
Betty Barrett

Resilience and Safety

Resilience is often defined in terms of the ability to continue operations or recover a stable state after a major mishap or event. This definition focuses on the reactive nature of resilience and the ability to recover after an upset. In this chapter, we use a more general definition that includes prevention of upsets. In our conception, resilience is the ability of systems to prevent or adapt to changing conditions in order to maintain (control over) a system property. In this chapter, the property we are concerned about is safety or risk. To ensure safety, the system or organization must be resilient in terms of avoiding failures and losses, as well as responding appropriately after the fact.

Major accidents are usually preceded by periods where the organization drifts toward states of increasing risk until the events occur that lead to a loss (Rasmussen, 1997). Our goal is to determine how to design resilient systems that respond to the pressures and influences causing the drift to states of higher risk or, if that is not possible, to design continuous risk management systems to detect the drift and assist in formulating appropriate responses before the loss event occurs.

Our approach rests on modeling and analyzing socio-technical systems and using the information gained in designing the socio-technical system, in evaluating both planned responses to events and suggested organizational policies to prevent adverse organizational drift, and in defining appropriate metrics to detect changes in risk (the equivalent of a 'canary in the coal mine'). To be useful, such modeling and analysis must be able to handle complex, tightly coupled systems with distributed human and automated control, advanced technology and software-intensive systems, and the organizational and social aspects of systems. To do this, we use a new model of accident causation (STAMP) based on system theory. STAMP includes non-linear, indirect, and feedback relationships and can better handle the levels of complexity and technological innovation in today's systems than traditional causality and accident models.

In the next section, we briefly describe STAMP. Then we show how STAMP models can be used to design and analyze resilience by applying it to the safety culture of the NASA Space Shuttle program.

STAMP

The approach we use rests on a new way of thinking about accidents, called STAMP or Systems-Theoretic Accident Modeling and Processes (Leveson, 2004), that integrates all aspects of risk, including organizational and social aspects. STAMP can be used as a foundation for new and improved approaches to accident investigation and analysis, hazard analysis and accident prevention, risk assessment and risk management, and for devising risk metrics and performance monitoring. In this chapter, we will concentrate on its uses for risk assessment and management and its relationship to system resilience. One unique aspect of this approach to risk management is the emphasis on the use of visualization and building shared mental models of complex system behavior among those responsible for managing risk. The techniques integral to STAMP can have great value in terms of more effective organizational decision-making.

Systems are viewed in STAMP as interrelated components that are kept in a state of dynamic equilibrium by feedback loops of information and control. A socio-technical system is not treated as just a static design, but as a dynamic process that is continually adapting to achieve its ends and to react to changes in itself and its environment. The original design must not only enforce constraints on behavior to ensure safe operations, but it must continue to operate safely as changes and adaptations occur over time.

Safety is an emergent system property. In STAMP, accidents are accordingly viewed as the result of flawed processes involving interactions among people, societal and organizational structures, engineering activities, and physical system components. The process leading up to an accident can be described in terms of an adaptive feedback function that fails to maintain safety as performance changes over time to meet a complex set of goals and values. The accident or loss itself results not simply from component failure (which is treated as a symptom of the problems) but from inadequate control of safety-related constraints on the development, design, construction, and operation of the socio-technical system.

Safety and resilience in this model are treated as *control* problems: Accidents occur when component failures, external disturbances, and/or dysfunctional interactions among system components are not adequately handled. In the Space Shuttle Challenger accident, for example, the O-rings did not adequately control the propellant gas release by sealing a tiny gap in the field joint. In the Mars Polar Lander loss, the software did not adequately control the descent speed of the spacecraft – it misinterpreted noise from a Hall effect sensor as an indication the spacecraft had reached the surface of the planet.

Accidents such as these, involving engineering design errors, may in turn stem from inadequate control of the development process, i.e., risk is not adequately managed in design, implementation, and manufacturing. Control is also imposed by the management functions in an organization – the Challenger and Columbia accidents, for example, involved inadequate controls in the launch-decision process and in the response to external pressures – and by the social and political system within which the organization exists. A system is resilient when that control function responds appropriately to prevent risk from increasing to unacceptable levels as the system and its environment change over time.

While events reflect the *effects* of dysfunctional interactions and inadequate enforcement of safety constraints, the inadequate control itself is only indirectly reflected by the events – the events are the *result* of the inadequate control. The control structure itself, therefore, must be carefully designed and evaluated to ensure that the controls are adequate to maintain the constraints on behavior necessary to control risk. This definition of risk management is broader than definitions that define it in terms of particular activities or tools. STAMP, which is based on systems and control theory, provides the theoretical foundation to develop the techniques and tools, including modeling tools, to assist managers in managing risk and building resilient systems in this broad context.

Note that the use of the term 'control' does not imply a strict military command and control structure. Behavior is controlled or influenced not only by direct management intervention, but also indirectly by policies, procedures, shared values, the level of awareness of behavioral/cognitive patterns and other aspects of the organizational culture. All behavior is influenced and at least partially 'controlled' by the social and organizational context in which the behavior occurs. Engineering this context can be an effective way of creating and changing a safety culture.

STAMP is constructed from three fundamental concepts: constraints, hierarchical levels of control, and process models. These concepts, in turn, give rise to a classification of control flaws that can lead to accidents. Each of these is described only briefly here; for more information see Leveson (2004).

The most basic component of STAMP is not an event, but a constraint. In systems theory and control theory, systems are viewed as hierarchical structures where each level imposes constraints on the activity of the level below it – that is, constraints or lack of constraints at a higher level allow or control lower-level behavior.

Safety-related constraints specify those relationships among system variables that constitute the non-hazardous or safe system states – for example, the power must never be on when the access to the high-voltage power source is open, the descent engines on the lander must remain on until the spacecraft reaches the planet surface, and two aircraft must never violate minimum separation requirements.

Instead of viewing accidents as the result of an initiating (root cause) event in a chain of events leading to a loss, accidents are viewed as resulting from interactions among components that violate the system safety constraints. The control processes that enforce these constraints must limit system behavior to the safe changes and adaptations implied by the constraints. Preventing accidents requires designing a control structure, encompassing the entire socio-technical system, that will enforce the necessary constraints on development and operations. Figure 8.1 shows a generic hierarchical safety control structure. Accidents result from inadequate enforcement of constraints on behavior (e.g., the physical system, engineering design, management, and regulatory behavior) at each level of the socio-technical system. Inadequate control may result from missing safety constraints, inadequately communicated constraints, or from constraints that are not enforced correctly at a lower level. Feedback during operations is critical here. For example, the safety analysis process that generates constraints always involves some basic assumptions about the operating environment of the process. When the environment changes such that those assumptions are no longer true, the controls in place may become inadequate.

The model in Figure 8.1 has two basic hierarchical control structures – one for system development (on the left) and one for system operation (on the right) – with interactions between them. A spacecraft manufacturer, for example, might only have system development under its immediate control, but safety involves both development and operational use of the spacecraft, and neither can be accomplished successfully in isolation: Safety must be designed into the physical system, and safety during operation depends partly on the original system design and partly on effective control over operations. Manufacturers must communicate to their customers the assumptions about the operational environment upon which their safety analysis and design was based, as well as information about safe operating procedures. The operational environment, in turn, provides feedback to the manufacturer about the performance of the system during operations.

Between the hierarchical levels of each control structure, effective communication channels are needed, both a downward *reference* channel providing the information necessary to impose constraints on the level below and a *measuring* channel to provide feedback about how effectively the constraints were enforced. For example, company management in the development process structure may provide a safety policy, standards, and resources to project management and in return receive status reports, risk assessment, and incident reports as feedback about the status of the project with respect to the safety constraints.

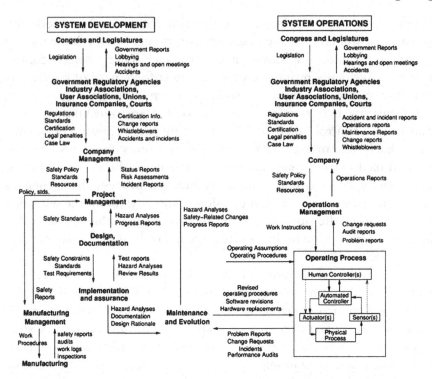

Figure 8.1: General form of a model of socio-technical control (Adapted from Leveson, 2004)

Between the hierarchical levels of each control structure, effective communication channels are needed, both a downward *reference* channel providing the information necessary to impose constraints on the level below and a *measuring* channel to provide feedback about how effectively the constraints were enforced. For example, company management in the development process structure may provide a safety policy, standards, and resources to project management and in return receive status reports, risk assessment, and incident reports as feedback about the status of the project with respect to the safety constraints.

The safety control structure often changes over time, which accounts for the observation that accidents in complex systems frequently involve a migration of the system toward a state where a small deviation (in the physical system or in human behavior) can lead to a catastrophe. The foundation for an accident is often laid years before. One event may trigger the loss, but if that event had not happened, another one would have. As an example, Figures 8.2 and 8.3 show the changes over time that led to a water contamination accident in Canada where 2400 people became ill and seven died (most of them children) (Leveson et al., 2003). The reasons why this accident occurred would take too many pages to explain and only a small part of the overall STAMP model is shown.

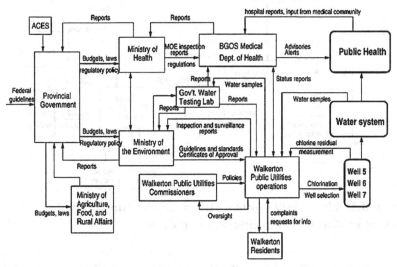

Figure 8.2: The designed safety control structure in the Walkerton Water contamination accident

Each component of the water quality control structure played a role in the accident. Figure 8.2 shows the control structure for water quality in Ontario Canada as designed. Figure 8.3 shows the control structure as it existed at the time of the accident. One of the important changes that contributed to the accident is the elimination of a government water-testing laboratory. The private companies that were substituted were not required to report instances of bacterial contamination to the appropriate government ministries. Essentially, the elimination of the feedback loops made it impossible for the government agencies and public utility managers to perform their oversight duties effectively. Note that the goal here is not to identify individuals to blame for the accident but to understand why they made the mistakes they made (none were evil or wanted children to die) and what changes are needed in the culture and water quality control structure to reduce risk in the future.

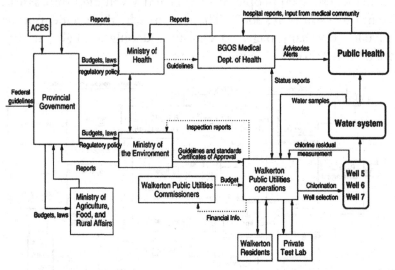

Figure 8.3: The safety control structure existing at the time of the Walkerton Water contamination accident

In this accident, and in most accidents, degradation in the safety margin occurred over time and without any particular single decision to do so but simply as a series of decisions that individually seemed safe but together resulted in moving the water quality control system structure slowly toward a situation where any slight error would lead to a major accident. Designing a resilient system requires ensuring that controls do not degrade despite the inevitable changes that occur over time or that such degradation is detected and corrected before a loss occurs.

Figures 8.2 and 8.3 show static models of the safety control structure, drawn in the form commonly used for control loops. (Lines going into the left of a box are control lines. Lines from or to the top or bottom of a box represent information, feedback, or a physical flow. Rectangles with sharp corners are controllers while rectangles with rounded corners represent physical processes.) But to build a resilient system, models are needed to understand *why* the structure changed over time in order to build in protection against unsafe changes. For this goal, we use system dynamics models. The field of system dynamics, created at MIT in the 1950s by Forrester, is designed to help decision makers learn about the structure and dynamics of complex systems, to design high leverage policies for sustained improvement, and to catalyze successful implementation and change. System dynamics provides a framework for dealing with dynamic complexity, where cause and effect are not obviously related. Like the other STAMP models, it is grounded in the theory of non-linear dynamics and feedback control, but also draws on cognitive and social psychology, organization theory, economics, and other social sciences (Sterman, 2000). System dynamics models are formal and can be executed, like our other models.

System dynamics is particularly relevant for complex systems. System dynamics makes it possible, for example, to understand and predict instances of policy resistance or the tendency for well-intentioned interventions to be defeated by the response of the system to the intervention itself. In related but separate research, Marais and Leveson are working on defining archetypical system dynamics models often associated with accidents to assist in creating the models for specific systems (Marais & Leveson, 2003).

Figure 8.4 shows a simple systems dynamics model of the Columbia accident. This model is only a hint of what a complete model might contain. The loops in the figure represent feedback control loops where the '+' or '-' on the loops represent polarity or the relationship (positive or negative) between state variables: a positive polarity means that the variables move in the same direction while a negative polarity means that they move in opposite directions. There are three main variables in the model: safety, complacency, and success in meeting launch rate expectations.

The control loop in the lower left corner of Figure 8.4, labeled R1 or *Pushing the Limit*, shows how, as external pressures increased, performance pressure increased, which led to increased launch rates and thus success in meeting the launch rate expectations, which in turn led to increased expectations and increasing performance pressures. This, of course, is an unstable system and cannot be maintained indefinitely – note the larger control loop, B1, in which this loop is embedded, is labeled *Limits to Success*. The upper left loop represents part of the safety program loop. The external influences of budget cuts and increasing performance pressures that reduced the priority of safety procedures led to a decrease in system safety efforts. The combination of this decrease along with loop B2 in which fixing problems increased complacency, which also contributed to reduction of system safety efforts, eventually led to a situation of (unrecognized) high risk. One thing not shown in the diagram is that these models can also contain delays. While reduction in safety efforts and lower prioritization of safety concerns may lead to accidents, accidents usually do not occur for a while, so false confidence is created that the reductions are having no impact on safety and therefore pressures increase to reduce the efforts and priority even further as the external performance pressures mount.

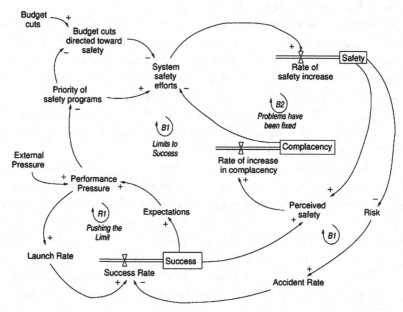

Figure 8.4: Simplified model of the dynamics behind the shuttle Columbia loss

The models can be used to devise and validate fixes for the problems and to design systems to be more resilient. For example, one way to eliminate the instability of the model in Figure 8.4 is to anchor the safety efforts by, perhaps, externally enforcing standards in order to prevent schedule and budget pressures from leading to reductions in the safety program. Other solutions are also possible. Alternatives can be evaluated for their potential effects and impact on system resilience using a more complete system dynamics model, as described in the next section.

Often, degradation of the control structure involves *asynchronous evolution*, where one part of a system changes without the related necessary changes in other parts. Changes to subsystems may be carefully designed, but consideration of their effects on other parts of the system, including the control aspects, may be neglected or inadequate. Asynchronous evolution may also occur when one part of a properly designed system deteriorates. The Ariane 5 trajectory changed from that of the Ariane 4, but the inertial reference system software did not. One factor in the loss of contact with the SOHO (Solar Heliospheric Observatory) spacecraft in 1998 was the failure to communicate to operators that a functional change had been made in a procedure to perform gyro spin-down.

Besides constraints and hierarchical levels of control, a third basic concept in STAMP is that of process models. *Any* controller – human or automated – must contain a model of the system being controlled. For humans, this model is generally referred to as their mental model of the process being controlled, cf. Figure 8.5.

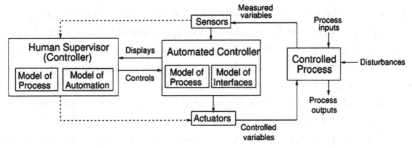

Figure 8.5: Role of models in control

For effective control, the process models must contain the following: (1) the current state of the system being controlled, (2) the required relationship between system variables, and (3) the ways the process can change state. Accidents, particularly system accidents, frequently result from inconsistencies between the model of the process used by the controllers and the actual process state; for example, the lander software thinks the lander has reached the surface and shuts down the descent engine; the Minister of Health has received no reports about water quality problems and believes the state of water quality in the town is better than it actually is; or a mission manager believes that foam shedding is a maintenance or turnaround issue only. Part of our modeling efforts involves creating the process models, examining the ways that they can become inconsistent with the actual state (e.g., missing or incorrect feedback), and determining what feedback loops are necessary to maintain the safety constraints.

When there are multiple controllers and decision makers, system accidents may also involve inadequate control actions and unexpected side effects of decisions or actions, again often the result of inconsistent process models. For example, two controllers may both think the other is making the required control action, or they make control actions that conflict with each other. Communication plays an important role here. Leplat suggests that accidents are most likely in *boundary* or *overlap* areas where two or more controllers control the same process (Leplat, 1987).

A STAMP modeling and analysis effort involves creating a model of the organizational safety structure including the static safety control structure and the safety constraints that each component is responsible for maintaining, process models representing the view of the process by those controlling it, and a model of the dynamics and pressures that can lead to degradation of this structure over time. These models and analysis procedures can be used to investigate accidents and incidents to determine the role played by the different components of the safety control structure and to learn how to prevent related accidents in the future, to proactively perform hazard analysis and design to reduce risk throughout the life of the system, and to support a continuous risk management program where risk is monitored and controlled.

In this chapter, we are concerned with resilience and therefore will concentrate on how system dynamics models can be used to design and analyze resilience, to evaluate the effect of potential policy changes on risk, and to create metrics and other performance measures to identify when risk is increasing to unacceptable levels. We demonstrate their use by modeling and analysis of the safety culture of the NASA Space Shuttle program and its impact on risk. The CAIB report noted that culture was a large component of the Columbia accident. The same point was made in the Rogers Commission Report on the Challenger

Accident (Rogers, 1987), although the cultural aspects of the accident was emphasized less in that report.

The models were constructed using both our personal long-term association with the NASA manned space program as well as interviews with current and former employees, books on NASA's safety culture such as McCurdy (1993), books on the Challenger and Columbia accidents, NASA mishap reports (Gehman, 2003), Mars Polar Lander (Young, 2000), Mars Climate Orbiter (Stephenson, 1999), WIRE (Branscome, 1999), SOHO (NASA/ESA Investigation Board, 1998), Huygens (Link, 2000), etc., other NASA reports on the manned space program (SIAT or Shuttle Independent Assessment Team Report (McDonald, 2000), and others) as well as many of the better researched magazine and newspaper articles.

We first describe system dynamics in more detail and then describe our models and examples of analyses that can be derived using them. We conclude with a general description of the implications for building and operating resilient systems.

The Models

System behavior in system dynamics is modeled by using feedback (causal) loops, stock and flows (levels and rates), and the non-linearities created by interactions among system components. In this view of the world, behavior over time (the dynamics of the system) can be explained by the interaction of positive and negative feedback loops (Senge, 1990). The models are constructed from three basic building blocks: positive feedback or reinforcing loops, negative feedback or balancing loops, and delays. Positive loops (called reinforcing loops) are self-reinforcing while negative loops tend to counteract change. Delays introduce potential instability into the system.

Figure 8.6a shows a *reinforcing loop*, which is a structure that feeds on itself to produce growth or decline. Reinforcing loops correspond to positive feedback loops in control theory. An increase in variable 1 leads to an increase in variable 2 (as indicated by the '+' sign), which leads to an increase in variable 1 and so on. The '+' does not mean the values necessarily increase, only that variable 1 and variable 2 will change in the same direction. If variable 1 decreases, then variable 2 will decrease. A '-' indicates that the values change in opposite directions. In the absence of external influences, both variable 1 and variable 2 will clearly grow or decline exponentially. Reinforcing loops generate growth, amplify deviations, and reinforce change (Sterman, 2000).

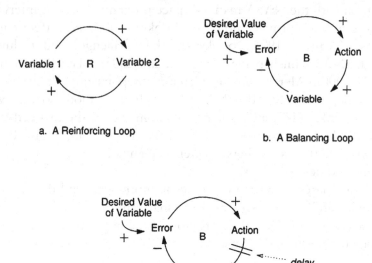

a. A Reinforcing Loop

b. A Balancing Loop

c. A Balancing Loop with a Delay

Figure 8.6: The three basic components of system dynamics models

A *balancing loop* (Figure 8.6b) is a structure that changes the current value of a system variable or a desired or reference variable through some action. It corresponds to a negative feedback loop in control theory. The difference between the current value and the desired value is perceived as an error. An action proportional to the error is taken to decrease the error so that, over time, the current value approaches the desired value.

The third basic element is a delay, which is used to model the time that elapses between cause and effect. A delay is indicated by a double line, as shown in Figure 8.6c. Delays make it difficult to link cause and effect (dynamic complexity) and may result in unstable system behavior. For example, in steering a ship there is a delay between a change in the rudder position and a corresponding course change, often leading to over-correction and instability.

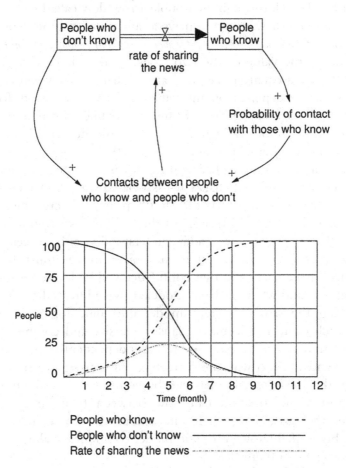

Figure 8.7: An example system dynamics and analysis output

The simple 'News Sharing' model in Figure 8.7 is helpful in understanding the stock and flow syntax and the results of our modeling effort. The model shows the flow of information through a population over time. The total population is fixed and includes 100 people. Initially, only one person knows the news, the other 99 people do not know it. Accordingly, there are two *stocks* in the model: *People who know* and *People who don't know*. The initial value for the *People who know* stock is one and that for the *People who don't know* stock is 99. Once a person learns the news, he or she moves from the left-hand stock to the right-hand stock through the double arrow flow called *Rate of sharing the news*. The rate of sharing the news at any point in time depends on the number of *Contacts between people who know and people who don't*, which is function of the value of the two stocks at that time. This function uses a differential equation, i.e., the rate of change of a variable V, i.e., dV/dt, at time t depends on the value of $V(t)$. The results for each stock and variable as a function of time are obtained through numerical integration. The graph in Figure 8.7 shows the numerical simulation output for the number of people who know, the number of people who don't know, and the rate of sharing the news as a function of time.

One of the significant challenges associated with modeling a socio-technical system as complex as the Shuttle program is creating a model that captures the critical intricacies of the real-life system, but is not so complex that it cannot be readily understood. To be accepted and therefore useful to risk decision makers, a model must have the confidence of the users and that confidence will be limited if the users cannot understand what has been modeled. We addressed this problem by breaking the overall system model into nine logical subsystem models, each of an intellectually manageable size and complexity. The subsystem models can be built and tested independently and then, after validation and comfort with the correctness of each subsystem model is achieved, the subsystem models can be connected to one another so that important information can flow between them and emergent properties that arise from their interactions can be included in the analysis. Figure 8.8 shows the nine model components along with the interactions among them.

As an example, our Launch Rate model uses a number of internal factors to determine the frequency at which the Shuttle can be launched. That value – the 'output' of the Launch Rate model – is then used by many other subsystem models including the Risk model and the Perceived Success by Management models.

The nine subsystem models are:

- Risk
- System Safety Resource Allocation
- System Safety Status

- System Safety Knowledge Skills and Staffing
- Shuttle Aging and Maintenance
- Launch Rate
- System Safety Efforts and Efficacy
- Incident Learning and Corrective Action
- Perceived Success by Management.

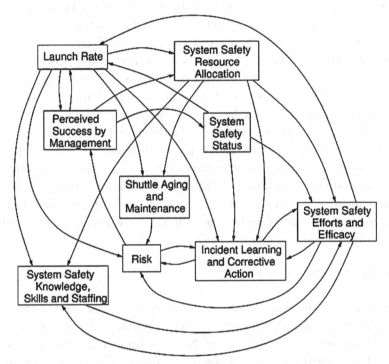

Figure 8.8: The nine submodels and their interactions

Each of these submodels is described briefly below, including both the outputs of the submodel and the factors used to determine the results. The models themselves can be found elsewhere (Leveson et al., 2005).

Risk

The purpose of the technical risk model is to determine the level of occurrence of anomalies and hazardous events, as well as the interval between accidents. The assumption behind the risk formulation is that once the system has reached a state of high risk, it is highly vulnerable to small deviations that can cascade into major accidents. The primary factors affecting the technical risk of the system are the effective age of the Shuttle, the quantity and quality of inspections aimed at uncovering and correcting safety problems, and the proactive hazard analysis and mitigation efforts used to continuously improve the safety of the system. Another factor affecting risk is the response of the program to anomalies and hazardous events (and, of course, mishaps or accidents).

The response to anomalies, hazardous events, and mishaps can either address the symptoms of the underlying problem or the root causes of the problems. Corrective actions that address the symptoms of a problem have insignificant effect on the technical risk and merely allow the system to continue operating while the underlying problems remain unresolved. On the other hand, corrective actions that address the root cause of a problem have a significant and lasting positive effect on reducing the system technical risk.

System Safety Resource Allocation

The purpose of the resource allocation model is to determine the level of resources allocated to system safety. To do this, we model the factors determining the portion of NASA's budget devoted to system safety. The critical factors here are the priority of the safety programs relative to other competing priorities such as launch performance and NASA safety history. The model assumes that if performance expectations are high or schedule pressure is tight, safety funding will decrease, particularly if NASA has had past safe operations.

System Safety Status

The safety organization's status plays an important role throughout the model, particularly in determining effectiveness in attracting high-quality employees and determining the likelihood of other employees becoming involved in the system safety process. Additionally, the status of the safety organization plays an important role in determining their level of influence, which in turn, contributes to the overall effectiveness of the safety activities. Management prioritization of system safety efforts plays an important role in this submodel, which in turn influences such safety culture factors as the power and authority of the safety organization, resource allocation, and rewards and recognition for raising safety concerns and placing emphasis on safety. In the model, the status of the safety organization has an impact on the ability to attract highly capable personnel; on the level of morale, motivation, and influence; and on the amount and effectiveness of cross-boundary communication.

System Safety Knowledge, Skills, and Staffing

The purpose of this submodel is to determine both the overall level of knowledge and skill in the system safety organization and to determine if the number of NASA system safety engineers is sufficient to oversee the contractors. These two values are used by the System Safety Effort and Efficacy submodel.

In order to determine these key values, the model tracks four quantities: the number of NASA employees working in system safety, the number of contractor system safety employees, the aggregate experience of the NASA employees, and the aggregate experience of the system safety contractors such as those working for United Space Alliance (USA) and other major Shuttle contractors.

The staffing numbers rise and fall based on the hiring, firing, attrition, and transfer rates of the employees and contractors. These rates are determined by several factors, including the amount of safety funding allocated, the portion of work to be contracted out, the age of NASA employees, and the stability of funding.

The amount of experience of the NASA and contractor system safety engineers relates to the new staff hiring rate and the quality of that staff. An organization that highly values safety will be able to attract better employees who bring more experience and can learn faster than lower quality staff. The rate at which the staff gains experience is also determined by training, performance feedback, and the workload they face.

Shuttle Aging and Maintenance

The age of the Shuttle and the amount of maintenance, refurbishments, and safety upgrades affects the technical risk of the system and the number of anomalies and hazardous events. The effective Shuttle age is mainly influenced by the launch rate. A higher launch rate will accelerate the aging of the Shuttle unless extensive maintenance and refurbishment are performed. The amount of maintenance depends on the resources available for maintenance at any given time. As the system ages, more maintenance may be required; if the resources devoted to maintenance are not adjusted accordingly, accelerated aging will occur.

The original design of the system also affects the maintenance requirements. Many compromises were made during the initial phase of the Shuttle design, trading off lower development costs for higher operations costs. Our model includes the original level of design for maintainability, which allows the investigation of scenarios during the analysis where system maintainability would have been a high priority from the beginning.

While launch rate and maintenance affect the rate of Shuttle aging, refurbishment and upgrades *decrease* the effective aging by providing complete replacements and upgrade of Shuttle systems such as avionics, fuel systems, and structural components. The amount of upgrades and refurbishment depends on the resources available, as well as on the perception of the remaining life of the system. Upgrades and refurbishment will most likely be delayed or canceled when there is high uncertainty about the remaining operating life. Uncertainty will be higher as the system approaches or exceeds its original design lifetime, especially if there is no clear vision and plan about the future of the manned space program.

Launch Rate

The Launch Rate submodel is at the core of the integrated model. Launch rate affects many parts of the model, such as the perception of the level of success achieved by the Shuttle program. A high launch rate without accidents contributes to the perception that the program is safe, eventually eroding the priority of system safety efforts. A high launch rate also accelerates system aging and creates schedule pressure, which hinders the ability of engineers to perform thorough problem investigation and to implement effective corrective actions that address the root cause of the problems rather than just the symptoms.

The launch rate in the model is largely determined by three factors:

1. Expectations from high-level management: Launch expectations will most likely be high if the program has been successful in the recent past. The expectations are reinforced through a 'Pushing the Limits' phenomenon where administrators expect ever more from a successful program, without necessarily providing the resources required to increase launch rate;

2. Schedule pressure from the backlog of flights scheduled: This backlog is affected by the launch commitments, which depend on factors such as ISS commitments, Hubble servicing requirements, and other scientific mission constraints;

3. Launch delays that may be caused by unanticipated safety problems: The number of launch delays depends on the technical risk, on the ability of system safety to uncover problems requiring launch delays, and on the power and authority of system safety personnel to delay launches.

System Safety Efforts and Efficacy

This submodel captures the effectiveness of system safety at identifying, tracking, and mitigating Shuttle system hazards. The success of these activities will affect the number of hazardous events and problems identified, as well as the quality and thoroughness of the resulting investigations and corrective actions. In the model, a combination of reactive problem investigation and proactive hazard mitigation efforts leads to effective safety-related decision-making that reduces the technical risk associated with the operation of the Shuttle. While effective system safety activities will improve safety over the long run, they may also result in a decreased launch rate over the short term by delaying launches when serious safety problems are identified.

The efficacy of the system safety activities depends on various factors. Some of these factors are defined outside this submodel, such as the availability of resources to be allocated to safety and the availability and effectiveness of safety processes and standards. Others depend on the characteristics of the system safety personnel themselves, such as their number, knowledge, experience, skills, motivation, and commitment. These personal characteristics also affect the ability of NASA to oversee and integrate the safety efforts of contractors, which is one dimension of system safety effectiveness. The quantity and quality of lessons learned and the ability of the organization to absorb and use these lessons is also a key component of system safety effectiveness.

Incident Learning and Corrective Action

The objective of this submodel is to capture the dynamics associated with the handling and resolution of safety-related anomalies and hazardous events. It is one of the most complex submodels, reflecting the complexity of the cognitive and behavioral processes involved in identifying, reporting, investigating, and resolving safety issues. Once integrated into the combined model, the amount and quality of learning achieved through the investigation and resolution of safety problems impacts the effectiveness of system safety efforts and the quality of resulting corrective actions, which in turn has a significant effect on the technical risk of the system.

The structure of this model revolves around the processing of incidents or hazardous events, from their initial identification to their eventual resolution. The number of safety-related incidents is a function of the technical risk. Some safety-related problems will be reported while others will be left unreported. The fraction of safety problems reported depends on the effectiveness of the reporting process, the employee sensitization to safety problems, and the possible fear of reporting if the organization discourages it, perhaps due to the impact on the schedule. Problem reporting will increase if employees see that their concerns are considered and acted upon, that is, if they have previous experience that reporting problems led to positive actions. The number of reported problems also varies as a function of the perceived safety of the system by engineers and technical workers. A problem-reporting positive feedback loop creates more reporting as the perceived risk increases, which is influenced by the number of problems reported and addressed. Numerous studies have shown that the risk perceived by engineers and technical workers is different from high-level management perception of risk. In our model, high-level management and engineers use different cues to evaluate risk and safety, which results in very different assessments.

A fraction of the anomalies reported are investigated in the model. This fraction varies based on the resources available, the overall number of anomalies being investigated at any time, and the thoroughness of the investigation process. The period of time the investigation lasts will also depend on these same variables.

Once a hazard event or anomaly has been investigated, four outcomes are possible: (1) no action is taken to resolve the problem, (2) a corrective action is taken that only addresses the symptoms of the problem, (3) a corrective action is performed that addresses the root causes of the problem, and (4) the proposed corrective action is rejected, which results in further investigation until a more satisfactory solution is proposed. Many factors are used to determine which of these four possible outcomes results, including the resources available, the schedule pressure, the quality of hazardous event or anomaly investigation, the investigation and resolution process and reviews, and the effectiveness of system safety decision-making. As the organization goes through this ongoing process of problem identification, investigation, and resolution, some lessons are learned, which may be of variable quality depending on the investigation process and thoroughness. In our model, if the safety personnel and decision-makers have the capability and resources to extract and internalize high-quality lessons from the investigation process, their overall ability to identify and resolve problems and to do effective hazard mitigation will be enhanced.

Perceived Success by Management

The purpose of this submodel is to capture the dynamics behind the success of the Shuttle program as perceived by high-level management and NASA administration. The success perceived by high-level management is a major component of the Pushing the Limit reinforcing loop, where much will be expected from a highly successful program, creating even higher expectations and performance pressure. High perceived success also creates the impression that the system is inherently safe and can be considered operational, thus reducing the priority of safety, which affects resource allocation and system safety status. Two main factors contribute to the perception of success: the accumulation of successful launches positively influences the perceived success while the occurrence of accidents and mishaps have a strong negative influence.

Principal Findings and Anticipated Outcomes/Benefits

The models we constructed can be used in many ways, including understanding how and why accidents have occurred, testing and validating changes and new policies (including risk and vulnerability assessment of policy changes), learning which 'levers' have a significant and sustainable effect, and facilitating the identification and tracking of metrics to detect increasing risk. But in order to trust the models and the results from their analysis, the users need to be comfortable with the models and their accuracy.

We first validated each model individually, using (1) review by experts familiar with NASA and experts on safety culture in general and (2) execution of the models to determine whether the results were reasonable.

Once we were comfortable with the individual models, we ran the integrated model using baseline parameters. In the graphs that follow, Figures 8.9-8.13, the arrows on the x-axis (timeline) indicate when accidents occur during the model execution (simulation). Also, it should be noted that we are *not* doing risk *assessment*, i.e., quantitative or qualitative calculation of the likelihood or severity of an accident or mishap. Instead, we are doing risk *analysis*, i.e., trying to understand the static causal structure and dynamic behavior of risk or, in other words, identifying what technical and organizational factors contribute to the level of risk and their relative contribution to the risk level, both at a particular point in time and as the organizational and technical factors change over time.

The first example analysis of the baseline models evaluates the relative level of concern between safety and performance (Figure 8.9). In a world of fixed resources, decisions are usually made on the perception of relative importance in achieving overall (mission) goals. Immediately after an accident, the perceived importance of safety rises above performance concerns for a short time. But performance quickly becomes the dominant concern.

Accidents lead to a re-evaluation of NASA safety and performance priorities but only for a short time:

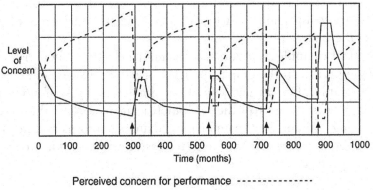

Perceived concern for performance ------------

Perceived concern for safety   ——————

Figure 8.9: Relative level of concern between safety and performance

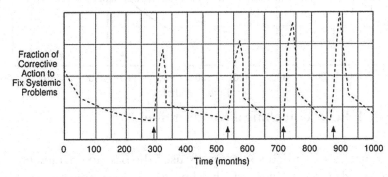

Attempts to fix systemic problems --------------

Figure 8.10: Fraction of corrective action to fix systemic safety problems over time

A second example looks at the fraction of corrective action to fix systemic safety problems over time (Figure 8.10). Note that after an accident, there is a lot of activity devoted to fixing systemic factors for a short time, but as shown in the previous graph, performance issues quickly dominate over safety efforts and less attention is paid to fixing the safety problems. The length of the period of high safety activity basically corresponds to the return to flight period. As soon as the Shuttle starts to fly again, performance becomes the major concern as shown in the first graph.

The final example examines the overall level of technical risk over time (Figure 8.11). In the graph, the level of risk decreases only slightly and temporarily after an accident. Over longer periods of time, risk continues to climb due to other risk-increasing factors in the models such as aging and deferred maintenance, fixing symptoms and not root causes, limited safety efforts due to resource allocation to other program aspects, etc.

Responses to accidents have little lasting impact on risk

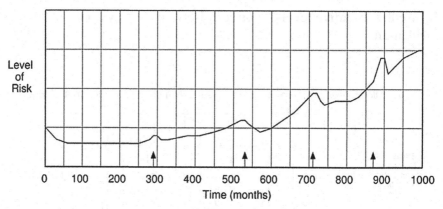

Figure 8.11: Level of technical risk over time

The analysis described so far simply used the baseline parameters in the integrated model. One of the important uses for our system dynamics models, however, is to determine the effect of changing those parameters. As the last part of our Phase 1 model construction and validation efforts, we ran scenarios that evaluated the impact of varying some of the model factors. Two examples are included here.

In the first example scenario, we examined the relative impact on level of risk from fixing symptoms only after an accident (e.g., foam shedding or O-ring design) versus fixing systemic factors (Figure 8.12). Risk quickly escalates if symptoms only are fixed and not the systemic factors involved in the accident. In the graph, the combination of fixing systemic factors and symptoms comes out worse than fixing only systemic factors because we assume a fixed amount of resources and therefore in the combined case only partial fixing of symptoms and systemic factors is accomplished.

Scenario 1: Impact of fixing systemic factors vs. symptoms

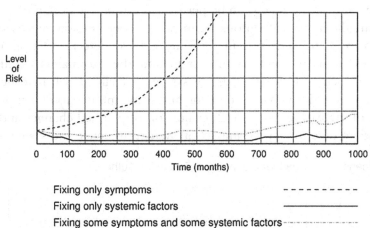

Fixing only symptoms   - - - - - - - - - -

Fixing only systemic factors   ——————

Fixing some symptoms and some systemic factors --·--·--·--·--·

Figure 8.12: Fixing symptoms vs. fixing systemic factors

A second example scenario looks at the impact on the model results of increasing the independence of safety decision makers through an organizational change like the Independent Technical Authority (Figure 8.13). The decreased level of risk arises from our assumptions that the ITA will involve:

- The assignment of high-ranked and highly regarded personnel as safety decision-makers;
- Increased power and authority of the safety decision-makers;
- The ability to report increased investigation of anomalies; and
- An unbiased evaluation of proposed corrective actions that emphasize solutions that address systemic factors.

Note that although the ITA reduces risk, risk still increases over time. This increase occurs due to other factors that tend to increase risk over time such as increasing complacency and Shuttle aging.

Implications for Designing and Operating Resilient Systems

We believe that the use of STAMP and, particularly, the use of system dynamics models can assist designers and engineers in building and operating more resilient systems. While the model-building activity described in this paper involved a retroactive analysis of an existing system, similar modeling can be performed during the design process to evaluate the impact of various technical and social design factors and to design more resilient systems that are able to detect and respond to changes in both the internal system and the external environment. During operations for an existing system, a resilient risk management program would involve (1) deriving and using metrics to detect drift to states of increasing risk and (2) evaluation of planned changes and new policies to determine their impact on system resilience and system risk.

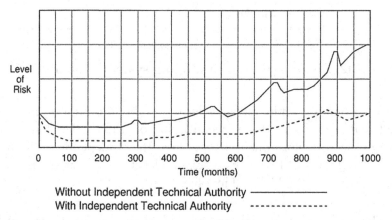

Scenario 2: Impact of Independent Technical Authority

Without Independent Technical Authority ——————
With Independent Technical Authority   - - - - - - - - - -

Figure 8.13: The impact of introducing an independent technical authority

The ability for evaluation of design factors and the heightened awareness of the critical shortcomings of an organizational management system or decision-making process is of real value and importance in the real world context that demands ever-increasing performance or productivity. The tension that managers and others in a organization feel as their deadlines are shortened, their resources constrained, and the competition intensifies often allows safety and risk considerations to diminish. Using STAMP and the models we introduce can potentially provide decision-makers with critical insights that can prevent organizations from drifting into less safe behaviors and promote greater adaptability. Organizations that operate in hazardous conditions must make every effort to have proper prevention as well as mitigation activities.

To evaluate the approach in a real setting, we are currently using it to perform a risk and vulnerability analysis of changes to the organizational structure of the NASA manned space program in the wake of the Columbia loss. We plan to follow this analysis up with a field study to determine how well the analysis predicted the risks over time and how accurate the modeling activity was in analyzing the resilience of the new NASA safety organization.

Chapter 9

Is Resilience Really Necessary?
The Case of Railways

Andrew Hale
Tom Heijer

Introduction

In other parts of this book resilience is discussed as a desirable attribute of organisations in managing safety in complex, high-risk environments. It seems to us necessary to explore also the boundaries of resilience as a strategy or state of organisations. Is it the only successful strategy in town? If not, is it always the most appropriate strategy? If not, when is it the preferred one? This chapter tries to address some of these questions, to give some thoughts and ways forward.

Railways as a Running Example

We take as our example the railways. We have been involved intensively in the last few years in a number of projects concerning safety management in the Dutch and European railways, partly sponsored by the European Union under the SAMRAIL and SAMNET projects (www.samnet.inrets.fr for official documents and reports of the project), and partly by the Dutch railway regulator (Hale et al., 2002). These have addressed the issues of the design of safety management systems for the increasingly privatised and decentralised railway systems in the member countries. In particular, our work has been concerned with the role of safety rules in risk control, and the methods of managing rules in the separate organisations and in the whole interacting railway system (Larsen et al., 2004; Hale et al., 2003). The information that we gathered from interviews, workshops and the study of documentation was designed to produce a best practice guideline for safety rule management. However, safety rules and how an organisation deals with them are such central issues for an organisation that we believe that the data can throw useful light on the question of organisational resilience. So we have recast the information we have and used it to reflect on the questions in this book. Inevitably there are gaps in what we can say, because the data were not originally collected to answer the questions posed here. However, we hope they prompt some thought and discussion and may even point out ways forward in both theory and practice.

The railway industry is particularly interesting in this context, since it prides itself, based on European Union figures (see Table 9.1) on being the safest form of transport in terms of passenger deaths per person kilometre (equal with aviation), or in deaths per person travel hour (equal with buses and coaches). In this sense it would like to see itself as an ultra-safe activity in Amalberti's terms (Amalberti, 2001). However, if we look at its record on track maintenance worker safety, its performance in the Netherlands is at or below the average of the construction industry in terms of deaths per employed person per year. The passenger train companies have a record of problems related to the safety of staff, particularly conductors and ticket controllers, where violent assaults by passengers resisting paying their fares are a problem that has been increasing in recent years. The railways are also one of the favourite mechanisms chosen by people in the Netherlands wishing to kill themselves (over 200 successful attempts per year) and it kills some 30 other transport participants per year at level crossings, as well as a number of trespassers.

These different safety performance indicators are looking at very different aspects of the railways and their management, some of which are only partly within the control of the participating companies, at least in the short term. For example, the risk of road traffic crossing railways can be greatly improved in the long term by eliminating level crossings in favour of bridges or underpasses. However, this would require, for such flat countries as the Netherlands, a very large investment over a long time. Although this is being actively worked on, it has to be coordinated with, and some of the money has to come from, the road managers. We will therefore concentrate here on the two aspects that are most within the control of the railway companies and their contractors, namely passenger safety and the safety of track maintenance workers. By looking at these we want to pose the question of whether railways are an example of a resilient system or not? Do they achieve their high level of passenger safety by using the characteristics that are typical of resilient organisations and systems, or is their success due to another strategy? Does the poor performance on track worker safety arise from a lack of resilience? The majority of the information comes from our Dutch study, but some is also used from the European cases.

Table 9.1: Comparative transport safety statistics 2001/2002 (European Transport Safety Council, 2003)

	Deaths/100 million person kilometres	Deaths/100 million travel hours
Road: total, of which:	0.95	28
Motorcycle/moped	13.8	440
Foot	6.4	25
Cycle	5.4	75
Car	0.7	25
Bus/coach	0.07	2
Ferry	0.25	8
Air (civil aviation)	0.035	16
Rail	0.035	2

Origins of the Data Used. The safety rule study was conducted using a protocol of questions based on the following framework of steps (Figure 9.1), which we expected to find carried out in some shape or form in the railway system, both within the different companies, and across the interfaces where they have to cooperate. The case studies concentrated on the maintenance of infrastructure, as it is one of the most complex and dangerous areas where safety rules play a big role.

The interviews were based on this framework and were conducted with a range of informants, who worked for the infrastructure company, a train operating company, several track maintenance companies, the traffic control and the regulatory bodies. The topics covered by the framework range over the link to the fundamental risk assessment process and the choice of preventive strategies and measures, before diving more deeply into the way in which rules and procedures are developed, approved, implemented and, above all, monitored, enforced or reviewed. The interviews therefore gave a broad coverage of many aspects of the safety management system and culture of the organisations concerned, as well as the communication and coordination across the organisational boundaries that make up the decentralised railway system.

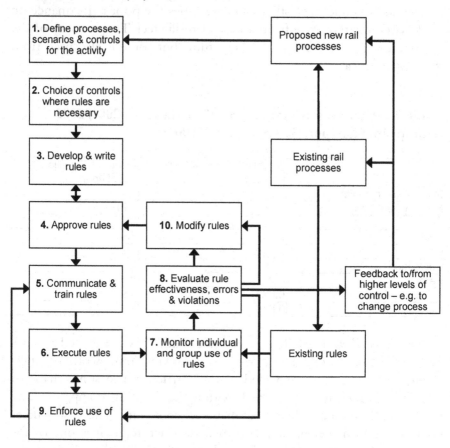

Figure 9.1: Framework for assessing the adequacy of management of safety procedures and rules (Larsen et al., 2004)

'Rules and procedures' were also interpreted broadly, to include everything from laws and regulations, through management procedures down to work instructions. We therefore ranged broadly in our interviews from policy makers in Ministries down to track maintenance supervisors. The observations that come out of the study can therefore be expected to give a good view of the resilience of the railway system.

A number of the researchers working on the Dutch and European projects were familiar with other high hazard industries, notably the chemical process, steel, nuclear and oil and gas industries. For a number of these researchers the SAMRAIL project was a first attempt to come to grips with the railway industry and its safety approaches and problems. The frame of reference they had, and which they expected to find in the railways, was conditioned by the very explicit requirements of the process industries. These included:

- a documented safety management system, subject to regular audit against extensive normative descriptions of what should be found in a well-managed system, and
- safety cases making explicit the major hazard scenarios and arguing that the levels of safety achieved by specified preventive measures, backed up by the safety management systems, are acceptable, often in comparison to quantitative risk criteria.

The study of the safety rules management system in railways produced a number of surprises, when compared with this background expectation. They are presented in the following section, largely in the form that they manifested themselves in the SAMRAIL study, notably in response to the questioning under boxes 1 and 2 and to a lesser extent boxes 6-8 in Figure 9.1. In the section on 'Assessing Resilience' we return to the insights presented, but now with spectacles of 'resilience' as it was defined at the workshop.

Observations on Safety Management in Railway Track Maintenance

Explicit Models of Accident Scenarios and Safety Measures

We were surprised to find that none of the railway organisations involved (the infrastructure manager, the train companies, but also not the regulator) could provide us with a clear overall picture of the accident scenarios they were trying to manage. Safety seemed to be embedded in the way in which the system ran and the personnel thought, in a way which appeared at first to resemble descriptions of generative organisations (Westrum, 1991; Hudson, 1999). The informal culture was imbued with safety and much informal communication took place about it. Yet there was no overall explicit model of how risk control was achieved. As part of our project, we had to develop a first set of generic scenarios, and barriers/measures used to control these scenarios, in order to analyse the part played by safety rules and procedures in controlling them. This situation is in marked contrast to what is found in the nuclear, oil and gas and chemical process industries (or in the US defence industry), where companies and regulators have worked for many years with explicit scenarios. These concentrated in the early years (1970s to 1990s) more, it is true, on the stage of the scenario after the loss of control or containment, rather than on the preventive stages before this occurred. However, prevention scenarios have become a standard part of the safety cases for these industries for many years.

A similar surprise had occurred when the Safety Advisory Committee at Schiphol airport was set up in 1997 in the wake of the Amsterdam Bijlmer El-Al Boeing crash (Hale, 2001). Requests to the different players at the airport (airlines, air traffic control, airport, customs, fuel and baggage handling, etc.) to supply an explicit description of the risks they had to control and how their safety management systems did this, notably how they interacted with each other in this control activity, led to long silences and blank looks, or statements that 'they were working on them'. Only some 8 years later is there significant progress in scenario and risk modelling at Schiphol, and this is still far short of what is desirable (VACS, 2003; Ale et al., 2004). A similar lack of an explicit model for air traffic control safety can be inferred from studies of that activity (Pilon et al., 2001; Drogoul et al., 2003). The Dutch railways would appear to be lagging even further behind in this respect.

Communication Channels

We were struck by the extreme centralisation of the system and the way it has consistently discouraged local communication, e.g., between train drivers and maintenance gangs on the tracks they are running on, and even direct (e.g., mobile telephone) communication between train drivers and traffic controllers. Communication is still seen as something that should occur primarily via the track signals set by the controllers.

Figure 9.2 is a diagram of the communication channels in the railway system at the time of our study. This shows the lack of fast and direct channels between different parts of the system. This paucity of communication channels, and the limited information available to the controllers (e.g., only which block a train is in and not where in that block and what it is doing – riding or stationary with a problem) means that the system operates to a considerable extent in an open-loop mode.

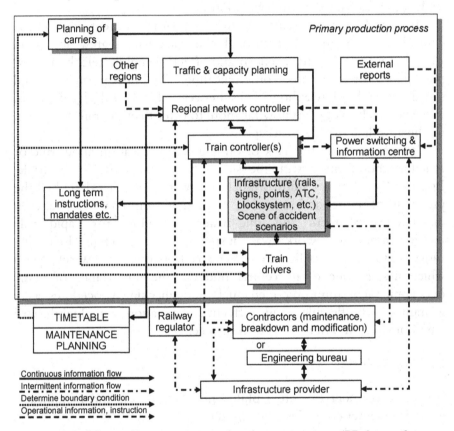

Figure 9.2: The railway communication structure (Hale et al., 2002)

The safety and other rules then take on a particular role as feedforward controls. They provide a degree of certainty about who will be doing what, where and when. Of course this certainty is only there if the rules and plans are strictly adhered to; hence the great emphasis on rule knowledge and obedience, which is characteristic of the railway system. We interpreted the discouragement of communication as the desire to stop rules being altered on-line outside the knowledge and approval of the controllers. Yet our interviews show widespread need for local adaptation, since many rules are locally impossible to apply as written. This is confirmed by many other studies of railway rules, e.g., Elling (1991), Free (1994), and Reason et al. (1998). The lack of communication channels to achieve this on-line adaptation is one factor leading to widespread routine violations.

It is perhaps understandable from a historic context why central control and communication through the track signals grew up. There were, before the advent of cheap and effective mobile telephony, few possibilities for distributed communication. The only option was fixed signals, with local telephone lines between distributed signal boxes. However, technology has now far more to offer than the railway system has been willing or able to accept.

We can contrast this lack of on-line communication about rule or schedule modification with the situation described by, for example, Bourrier (1998) in her study of nuclear power plants. She described the presence of close and effective communication channels between maintenance workers and rule-makers to allow for rapid rule modification. Her work was inspired by the High Reliability Organisation school at Berkeley (Weick & Roberts, 1993; Weick, 1995) which has, as one of its central findings, the importance of many overlapping communication channels and checking procedures as a guarantee of flexible and effective operation in high-hazard environments, in other words of resilience.

Operating Philosophy

When we consider operating philosophy, one of the most important issues is whether control is centralised or not.

Centralised Control. We concluded that this highly centralised method of control means that the system copes with threats to its process safety by operating an 'exclusive' regime. No train or work is officially allowed to proceed unless the requirements for safety are guaranteed to be present according to the rules. This is the essence of the block-working system common to all railways and developed after many years of trial and error involving train collisions (e.g., Holt, 1955). This assigns a defined section of the rail track to a single train and does not allow a following train (or opposing train on single line working) to enter it until the preceding train has left the block. This ensures, if all the rules are obeyed and nothing in the hardware fails to danger, that there is always a defined minimum distance between moving trains. Stop signs (signals) protect the block, and these must stay red until it is clear. The following train must wait at the red signal, even if it strongly suspects that the sign is defective. (This is in stark contrast to the behaviour of road traffic at traffic lights, where red lights, at least in the Netherlands, are treated by cyclists and pedestrians, and increasingly by drivers, as pretty coloured indicators to proceed with caution. The high level of road accidents may seem to show that this 'laxity' in road operations leads to low safety. However, the much higher traffic densities on the road should make us ask whether the relative safety is in fact lower – see Chapter 3 on definitions of resilience.) Only after communication with the controller, through the often-slow communication channels mentioned above, is permission given to proceed through a red signal. In other words, safety of the train in respect of avoidance of collisions is currently bought at the cost of a 'stop and restart' philosophy. If the system moves outside a certain expected and defined safe envelope, i.e., deviates from planned operation, the system has to be brought to a halt in parts at least, so that action can be taken to remove the threat, reposition the operating elements and then restart the system. No plans are made on a routine basis to cope with operation outside this envelope. The controllers therefore have to work out on the spot, or with a time horizon of only a few minutes, what to do in such cases. This often means that they are not able to do it fast enough to ensure smooth continuation of operations without introducing delays. This improvisation is in marked contrast to the planning of the normal timetable, which starts around a year before the operations take place. This approach can cost a great deal in punctuality of running, if things begin to slip in the timetable, due to unplanned delays, and trains have to wait at unprogrammed points before they can occupy critical parts of the network.

A future development, on which the railway industry has been working for many years, is to define the protected blocks not in terms of the fixed infrastructure, but in terms of a fixed envelope around a moving train. This could make the blocks smaller, but does not inherently increase the flexibility. The driving force to develop and use this technology is, therefore, to increase the capacity of the network, by allowing trains to run more closely one after the other.

Towards Decentralisation? Aviation has similar rules allocating slots and route segments to aircraft, particularly at airports and at busy parts of the network, but these are already related to the moving aircraft itself and not to fixed segments of the infrastructure. In this system there are also new developments, but these would represent a much greater revolution in operating philosophy, which might move that system much more towards resilience and flexibility. The 'free flight' concept (Hoekstra, 2001) would leave the resolution of conflicts between aircraft not to a central controller, as now, but to apparatus on board the individual aircraft. These would detect other aircraft with potentially conflicting trajectories and issue advisory flight path changes to the pilot. Hoekstra proposed a simple algorithm, which would ensure that the changes made by the two conflicting aircraft would always lessen the conflict and not cancel each other out. His simulation studies showed that this simple distributed system could cope effectively with traffic densities and conflict situations many times denser and more threatening than currently found in even the busiest airspace. Yet air traffic controllers, with their current central control philosophy, are already worried about their ability to cope with existing densities. Whilst there is still strong opposition to the free flight concept, such simulation studies do call into question the centralised control that both railways and aviation appear to be wedded to. See also Pariès' discussion (Chapter 4) of the possibly inherently safe characteristics of distributed systems.

Track Maintenance Safety

In contrast to the philosophy of passenger (= process) safety, the trade-offs for track maintenance seem to occur very differently. Here the maintenance teams have in the past been expected to 'dodge between the trains' to do their work. Only a minority of inspection and maintenance work is done with complete closure of the track to be maintained, let alone of the track running parallel to it a few metres away. The tracks are laid close together, particularly in the crowded Netherlands, in order to use up as little land as possible. A project manager for some new track being laid told us that he had requested permission to lay the parallel tracks further apart to allow physical barriers to be inserted between them when maintenance was being conducted on one track. This would avoid the need for evacuation of the maintenance area when trains passed on the adjacent line, because it would prevent the maintenance workers inadvertently straying onto, or being sucked onto the adjacent tracks and would reduce the risk of equipment in use on or around the one track breaching the safe envelope of the adjacent track. It would also reduce to some extent the need for very slow running of trains on the adjacent track. He was told that this proposal would only cost more money and that there was no decision-making or accounting mechanism in place to discount that extra cost against the savings from faster, safer and more effective maintenance in the later stage of the life cycle.

Controlled Access. There is, according to the Dutch railway rulebook, a regime of 'controlled access' to track under maintenance. This involves handing control of a section of track over to the maintenance team, who actively block the signal giving access to it and can make trains wait until they have cleared the track before admitting them to it. This can be done by various means, but the commonest is to place a specially designed 'lance' between the two rails to short circuit them and activate the upstream signal, which thinks there is a train in that block of track, until the lance is physically removed. Implementation of this regime would require some development of new technology and rules. Train controllers have resisted implementation of this regime. The conclusion we drew from our study was that their underlying reason for doing this is that it would undermine their central control of the network and would make their decisions about train running, especially in unplanned situations, dependent on others outside their direct control. Because this regime is hardly ever used, most maintenance work is done with lookouts warning the maintenance workers to clear the track at the approach of the train. Studies in different countries (e.g., Hale, 1989; Itoh et al., 2004) have shown that this system is subject to many opportunities for errors and violations, often in the name of getting the work done according to the tight maintenance schedules. Under this regime the casualties among workers are relatively high. Track capacity is bought here at the cost of safety.

Trading Risks. As a coda to this point it is interesting to relate that the Dutch labour inspectorate has begun in the last few years to insist on the closure of tracks and even adjacent tracks to protect maintenance workers (its constituents) and has come into conflict in that respect not only with the infrastructure manager (responsible for the track integrity, but also for allocating (selling) capacity to the train operators), but also with the railway inspectorate, which is concerned with passenger safety. From calculations made for the latter (Heijer & Hale, 2002), it is clear that, if the tracks were to be closed completely for maintenance and the trains cancelled, the increase of risk for the passengers by transferring them from rail to bus or car to cover the same journey would far outweigh the extra safety of the maintenance workers by closing the parallel line. This type of calculation was something new in the debate between the different parties, which had, up to then, argued purely from the point of view of the interests of their own constituents.

Assessing Resilience

The observations made above are by no means an exhaustive set of relevant observations about the resilience of the railway system. The data were not collected explicitly for that reason, which means there are many gaps in what we can say from them. It would be necessary at least to discover how risk controls are developed and used over time in order to be more certain about those aspects of how the system adapts itself to change. However, we can use these insights as a preliminary set of data and conclusions to suggest questions for further study and debate. What follows is, therefore, in many respects speculative.

Woods (2003) listed five characteristics as indicating lack of resilience in organisations. These were expanded on during the workshop. We have compiled the following list of topics from those two sources, on which to assess the railways, in so far as our data allows that.

- Defences erode under production pressure.
- Past good performance is taken as a reason for future confidence (complacency) about risk control.
- Fragmented problem-solving clouds the big picture – mindfulness is not based on a shared risk picture.

- There is a failure to revise risk assessments appropriately as new evidence accumulates.
- Breakdown at boundaries impedes communication and coordination, which do not have sufficient richness and redundancy.
- The organisation cannot respond flexibly to (rapidly) changing demands and is not able to cope with unexpected situations.
- There is not a high enough 'devotion' to safety above or alongside other system goals.
- Safety is not built as inherently as possible into the system and the way it operates, by default.

We might wish to question whether we need the term 'resilience' to define these characteristics, since many have already been discussed extensively in the literature on safety management and culture, which we have referenced above. However, we will take resilience as a working title for them. In the next sections we compare the information we collected from the railway industry with these characteristics and come to the conclusion that the industry scores relatively poorly. Hence we conclude that the current good passenger safety performance seems not to be achieved by resilience.

Defences Erode under Production Pressure

We have indicated above that we believe that the relatively poor record on track maintenance worker safety is related to the sacrifice of safety for the train timetable. There is strong resistance to the sort of track and line closures that would be needed to make a significant improvement in maintenance safety. There are also too few mechanisms for considering the needs of maintenance safety at the design stage of new track. In contrast, train and passenger safety shows the opposite trade-off at present, with delays in the timetable being accepted as the price of the stop-restart philosophy, which is needed to guarantee passenger safety. Currently the railways are resilient in respect of passenger safety according to this criterion, but not in respect of track maintenance worker safety.

It is important to ask whether this current resistance to production pressure in relation to passenger safety will be maintained. There has been intensive pressure the last five years on the Dutch railway system to improve its punctuality. Heavy fines have been imposed by government under performance contracts for falling below the targets set in relation to late running and outfall of trains. At the same time, the need for an extensive programme to recover from a long period of too limited track maintenance has been recognised and launched. This has resulted from and led to the discovery of many cracked rails and damaged track bedding. This requires very major increases in the maintenance program, to achieve its primary aim of sustaining and improving primary process safety. However, the capacity of railway contractors is limited and large infrastructure projects like the high speed line from Amsterdam to Brussels and the Betuwe Route (a dedicated new freight route from the Rotterdam harbours to Germany) occupy a large part of that capacity (and the available funds). Train controllers have been given targets linked more closely to these punctuality targets, as their task has been modified to give more emphasis to their function as allocators of network capacity and not just as guardians of its safety. This has led, according to our interviews, to increasing conflicts with train drivers reporting safety-related problems, which would require delays. All of the signs are therefore present for erosion of safety margins to take place. Our data cannot indicate whether or not such erosion is taking place. That would require a much more extensive analysis of subordinate indicators of safety, such as maintenance backlogs, violations of manning rules, or other signs of compromises between safety and the timetable being increasingly decided in favour of the timetable. The jury is therefore still out on this issue.

Past (Good) Performance is Taken as a Reason for Future Confidence –
Complacency

There are less clear signs here from our interviews. These showed that the personnel are proud of their passenger safety record and talk about that aspect of performance to the relative exclusion of the other less favourable indicators, such as the track maintenance worker safety, or issues such as level crossing safety. However, there is also widespread concern about safety performance and about the likelihood that it will deteriorate in the face of the privatisation and decentralisation, which is far advanced. So we did not find any widespread complacency, certainly about passenger safety. Track worker safety and level crossing safety are also priority areas identified in the annual report of the Dutch infrastructure manager (Prorail, 2003).

What we do find in documents and workshops (Dijkstra & v. Poortvliet, 2004) is extensive argumentation that still more demands for investment in passenger safety for rail transport would be counterproductive. They would be inclined to increase the prices of rail tickets, restrict the number and punctuality of trains, and hence discourage people from making the modal transfer from the road to the rail systems. We do not, however, interpret this as complacency about the continued achievement of the high levels of passenger safety. It seems to fit more into an active concern with system boundaries, deriving from just the sorts of analysis of relative safety between modes, which we indicated in the section on track maintenance safety above. In general, the railways would appear to be resilient according to this criterion.

Fragmented Problem-Solving Clouds the Big Picture. No Shared Risk Picture

We consider a number of topics under this heading: the presence of an explicit risk model for the railway system, whether an integrated view is taken of all types of safety (passengers, track workers, drivers at level crossings, trespassers, etc.) and the issue of privatisation and decentralisation as threats to an integrated picture of risk control.

No Explicit Risk Model. As indicated in the preceding discussion of models of accident scenarios and safety measures, the Dutch railway system does not seem to have an explicit big picture defined in relation to its risk control task. The requirement of a system controller that it has a good model and overview of the system to be controlled is certainly not explicitly met. When we looked at the implicit picture its managers and workforce have, as seen through the scenarios we defined and the control measures we found in place to prevent, recognise and recover from, or mitigate the scenarios, we found that pretty well all those we defined during normal operation, maintenance and emergency situations were covered by rules or built-in barriers, but that those during the transitions between those phases were not. For example, a large number of the serious accidents to maintenance personnel occur in the phase of getting to or from their workplace, which is relatively poorly covered by procedures and safety measures. We also found the same sort of controversy raging about safety regimes, which we have described in the discussion of track maintenance safety. This represents a different view of how safety should be achieved, between the traffic controllers and the contractors required to do the maintenance.

An Integrated View of all Types of Safety? A further indicator of a lack of a clear and agreed risk picture is a split between the departments responsible for the different types of safety in the system. This split used to be far deeper in railways (Hale, 1989), with completely different departments in the old monopoly company dealing with train and passenger safety and with the working safety of staff. Within the infrastructure manager this split has now been considerably reduced. The safety departments have been fused. Safety indicators are brought together and presented in annual reports (Prorail, 2003) under one heading. There is also a move in the last few years to relate and trade-off the different risk numbers quite explicitly, implying an overarching risk picture at the level of goals.

When we look at the translation of this increasing unity of goals into specific risk control measures, we see a different picture. The technical departments dealing with each type of risk are still relatively separate, as our example, in the section on track maintenance safety, of the lack of trade-off possible between track separation and maintenance costing shows. The consideration of different types of safety measures is also uneven and unintegrated. Relatively recent risk assessment studies of tunnel safety (Vernez et al., 1999) still show a relative overkill in the use of technical safety measures, with far too little integrated consideration of the concomitant organisational and human factors issues of clear responsibilities, ergonomic design, communication and (emergency) organisation. Our study of safety rule management (Larsen et al., 2004) showed that the railway system is still a technocratic one, with a mechanistic view of human behaviour and the role of rules in shaping that behaviour. There is little recognition of the need for on-line safety to be managed flexibly by those 'at the sharp end', which implies that rules and plans must be modifiable on-line to cater for local circumstances and situations. Instead rules are seen as carved in stone, whilst rule management systems often lack clear monitoring and modification steps, particularly those that would involve 'actors' now working in different organisations, but whose actions interact in achieving railway safety. We can characterise these safety measures as relatively rigid and lacking in resilience.

Decentralisation and Privatisation. With the privatisation and splitting up of the system which has taken place over the last few years, there is now extensive fragmentation of the system, both vertically in the trajectory from the regulator down through the railway undertakings to the contractors, and horizontally in the splitting of the old monopolies into separate infrastructure, capacity management and train operation companies. The old unity, which was created in the monopoly railway undertakings through the traditions of employment from father to son, through life-long employment and restricted use of external contracting, has now largely disappeared in the Netherlands. This unity resulted in a relatively common, if technocratic and rigid, picture of how safety was achieved. As we have indicated, that picture was largely implicit. With new train operating companies, maintenance contractors and sub-contractors now being employed, and with the increasing pressure for the opening of the national markets to foreign concessionaries, the basis for this common risk picture and practice is being rapidly eroded. National differences in safety practices are also being made apparent, as cross-border operations have increased and as the European drive towards inter-operability of railway systems has gathered momentum. Many of these are specific and different solutions to inherent and common problems, such as the form and content of signals and signs, the way in which rule violations are dealt with, or how control philosophies are translated into specific hardware requirements. The European TSIs (Technical Specifications for Interoperability), agreed at European level between representatives of national railways, are gradually reimposing a common risk picture, but this process will take a long time, as each country is required to change at least some of its current methods. Such a changeover period always produces the opportunity for misinterpretation and error.

Conclusion. The balance according to this criterion is that the old resilience, built into the system through its strong safety culture and nurtured by a stable workforce, is being rapidly eroded. Because the risk picture was never made explicit, it is particularly vulnerable to such cultural erosion. On the positive side, the railway industry has become increasingly open to the outside world over the last decade. This has meant that there is now much more attention to the need for a clear and explicit safety philosophy and safety management system. The SAMRAIL project is a manifestation of this. In that project there was explicit attention for the question of what the railway industry could learn from other high technology, high hazard industries, such as chemical process, nuclear and aviation.

Failure to Revise Risk Assessments as New Evidence Accumulates

Here our study does not offer any direct evidence. We have not penetrated deeply enough into the detailed workings of the system in our studies to follow the sort of risk assessment processes over time that would be necessary to draw conclusions on this point. The concerns about the still relatively watertight compartments within which decisions about risk are made, make us suspect that there is a certain rigidity in dealing with risks. Comparison of risks across domains is still a relatively new activity, meaning that priorities are not yet clearly open to risk-based thinking. However, the examples of normalisation of deviance analysed by Vaughan (1996) for Challenger and in the Columbia report (Gehman, 2003), which both used risk-based reasoning to justify risk acceptance, do not make us optimistic that such risk-based priorities will necessarily be safer ones. We already see risk comparisons being used for justifying not spending money on low risk problems (Dijkstra & v. Poortvliet, 2004) rather than for justifying using money on higher risks.

Breakdown at Boundaries Impeding Communication and Coordination

The break-up of the old rail monopolies mentioned in the discussion of fragmented problem solving above, has led to a significant decrease in formal communication and coordination. Our interviews showed that many existing formal communication and joint problem-solving meetings across what have become external organisational boundaries, rather than interdepartmental ones, had been discontinued; those which still existed had become poorly attended, with members complaining that their bosses had told them to stick to core business and not waste time on such meetings. Some channels had been diverted, e.g., a previous direct reporting channel to the train controllers for train drivers to report things that worried them during the trip (such as trespassers or objects on or near the line, rough track, anomalies in signals or level crossings) had been diverted to a 'Helpdesk', which seemed to function in the manner too often found with such 'animals' i.e., as a barrier and delaying mechanism, rather than as a problem-solving organ. We found that informal communication channels were still working, based on people who had previously worked together, and who still contacted each other to resolve these communication issues, even though they were now working for different organisations. This often occurred despite a lack of support from, and sometimes against the express instructions of their bosses. These channels are, of course, vulnerable to these individuals leaving their current jobs or retiring.

We did, however, find evidence of the development of new coordinating mechanisms, seemingly as a result of the falling away of the old ones. For example the large maintenance contractors, despite their competition with each other, had recently formed a group to write harmonised safety rules and procedures for their own employees and sub-contractors, since they no longer got those in an operational form from the infrastructure manager. This group was also starting to put pressure on the other players to restore the necessary communication to cope with on-line rule modification and to discuss the issue of regimes under which maintenance would take place (see the earlier discussions on track maintenance safety and on the fragmented risk picture).

If we add to these points the inherently difficult nature of communication and coordination in a highly distributed system, spread across the whole country, with staff constantly on the move, the verdict according to this criterion must be that railways are currently not very resilient and are threatening to become even less so.

The Organisation Cannot Respond Flexibly to Changing Demands and Unexpected Situations

We have discussed the issues relevant to this criterion in the section on operating philosophy above. The railway system scores very poorly on this point. Our thesis is that it can achieve passenger safety only through its 'stop and restart' philosophy, and fails to achieve track maintenance safety because it does not adopt the same philosophy there.

Another prerequisite for flexibility is that many people can do the same tasks and so take over from colleagues who are unable for any reason to perform them. Railways are generally poor at this. The reasons are partly structural – there is a lot of single manning, such as for train drivers, making substitution problematical without incurring very high duplication costs. This is demonstrated when a driver becomes incapacitated or emotionally stressed by, for example, a suicide jumping in front of the train.

Not a Great Enough Devotion to Safety

The conclusion from our study of safety rules (Larsen et al., 2004) was that there was still a very strong concern for passenger safety at all levels in the organisation, sustained by the still strong culture we have described above. Our thesis was that this enabled the railway system to retain its high passenger safety performance, despite the other shortcomings we have. Our interviews revealed a less fanatical devotion to maintenance safety, where accidents were regarded rather more as a fact of life. However, in recent years and under the growing influence of the Labour Inspectorate, increasing attention has been devoted to track worker safety (e.g., the Safe Working on Infrastructure project). This has resulted, amongst other things, in instructions to perform most maintenance during train-free periods and only stopping a train service if necessary.

Our concern for the devotion to passenger safety was one directed to the future. If the fragmentation and internationalisation of the railway industry waters down this strong culture, and if the increasing performance pressures we have identified discussed above erode the determination of senior managers to place safety first, we can see that this strong safety culture could disappear, perhaps even at a rapid pace. Reorganisations, of which the system has had many in many countries in the last few years, are quite capable of destroying a good safety culture in a relatively short time, if turnover is high and job security low.

Against this trend we can identify the increasing importance of passenger safety as a core marketing value for railways. In an on-going project for a high-speed line operator, we see this being placed centrally at the highest level in the management system goals and performance criteria. However, we can also note that the overall responsibility for safety in the reorganised railway industry has been a subject of considerable lack of clarity, at least in the Netherlands. The position of the rail regulator within the Ministry, and/or within the infrastructure company was long unclear, undermining the quality of the regulatory process (Hale et al., 2002). The shifting priorities of the traffic controllers, from pure safety to more concern for punctuality, also chips away at one requirement identified in resilient companies, the presence of powerful gatekeepers whose only responsibility is the achievement of high levels of safety.

The large infrastructure projects that are currently being carried out in the Netherlands (e.g., the high speed line to Paris and the Betuwe goods line from the Rotterdam port to Germany), on the other hand, do have a much higher level of thinking about safety, which would bring those new systems to a higher, more comprehensive, level. There is some evidence that those insights will also influence safety considerations in the old railway system. In time this may well result in considerably higher resilience on this point.

Safety not Built Inherently into the System

Our study has not penetrated into sufficient depth to conduct a thorough analysis of this criterion for the operation of railways. The long history of trial and error learning discussed by such books as that by Holt (1955) suggests that the traditional operations have built in this safety very well over time. The very performance in relation to current passenger safety in Europe is testament to this. The current phase of major technological and organisational development (high-speed cross-border lines, new technical train management and safety system concepts, dynamic train management, mixed concepts of heavy and light rail, etc.) will provide a new opportunity to increase inherent safety, but also a new opportunity to make wrong choices. Whereas change management, particularly of technological change, is an integral part of safety management in the chemical industry, in order to guide change to the positive side, the railway industry has still to introduce that aspect into its safety management. We do not, therefore, draw any conclusions in this area in respect of resilience.

In the area of track maintenance work, however, we do draw a negative conclusion. The examples of failure to implement inherently safer maintenance regimes and to take account of maintenance safety at the design and layout stage discussed earlier are clear indications that the railway system is not resilient in respect of this aspect.

Discussion and Conclusions

Railways, on this assessment, would seem to be examples of poor, or at best mixed, resilience, which can, however, still achieve high levels of safety, at least in certain areas of their operations. Table 9.2 summarises the verdict on the eight criteria discussed on the last section.

Table 9.2: Summary assessment of resilience of railways

Criterion	Conclusion (+ resilient, - not)
1. Defences erode with production pressure	+/-. Mixed evidence & gaps
2. Past performance leads to complacency	+
3. No shared risk picture	Was +, is now becoming -
4. Risk assessment not revised with new evidence	? No data from our study
5. Boundary breakdown impedes communication	- and getting worse
6. No flexible response to change/unexpected	-
7. Not high enough devotion to safety	+, but under threat
8. Safety not inherent enough in design of system	-, but gaps in our data

Passenger safety appears to be achieved by defining very clearly in advance what are the necessary prerequisites of safe operation and forbidding operation outside them. This is an extreme version of a command and control strategy, which has been defined by the protagonists of high reliability organisations (Weick, 1995) as the opposite pole to the flexible, mindful organisation called in this workshop the 'resilient organisation'. When the system moves outside this clearly defined safe envelope, the railway system stops, regroups and restarts only when the necessary operating conditions have been re-established. Hence safety is achieved by sacrificing goals, traffic volume and punctuality. The system does not achieve all its goals simultaneously and flexibly and is not resilient. Yet it achieves a very high safety performance on this performance measure. We have argued that this level of performance may be under threat, because we see the current level of resilience tending to be reduced rather than increased.

On track maintenance worker safety, the system has a poorer performance. Here we can conclude that the lack of resilience on the criteria mentioned provide a reasonable summary of the reasons for this safety performance failure. The trade-off works here in the other direction. Punctuality of the trains and traffic volume are bought at the price of safety. The system is not resilient and performs relatively poorly. We fear, here also, a decline in performance as the resilience is reduced even further.

If our analysis is accepted, we have shown that safety performance and resilience are, at least here, relatively independent of each other. This complements the questions raised in our brief intervention (Chapter 3), in which we argued that road traffic could be called resilient in some senses, but performs in absolute terms (accidents per passenger kilometre) poorly. We said there that the performance was, however, quite good considering the exposure to risk.

We would appear to have, with railways, a system that is ultra-safe on one safety dimension, but not a good performer on several others. We could argue further along two lines:

- If railways were to work on improving their resilience, they would be able to become much better safety performers on other safety dimensions, without sacrificing other system goals. The large new projects like HSL and Betuwe route provide an opportunity for introduction of such improvements. We would then need to study if and how more resilience could be built in the conventional system and what that would cost.
- Resilience is only one strategy for achieving very high levels of safety. There are others, but these have other costs. We would then need to define when resilience is a suitable strategy for a company or system and when other strategies are 'better'.

We prefer to leave both of those lines of development open for further discussion, research and analysis.

Systems Are Never Perfect

Yushi Fujita

Engineers may well accomplish self-consistent design, but its outcome, the artifact, can never be perfect in operation. Neither humans working at the front end (e.g., operators, maintenance persons), nor humans working at the back end (e.g., administrators, regulators) are perfect. The system (i.e., the combination of artifact and humans with various responsibilities) therefore cannot be perfect. It is not only the variability of human performances but also the human propensity for pursuing perfection with prescriptive rules while ceaselessly trying to change the system for better, that makes the system incomplete and imperfect as the time passes by.

Chapter 10

Structure for Management of Weak and Diffuse Signals

Lars Axelsson

> *After the mid-16th century, samurai general Motonari Mohri had conquered a big part of Japan he feared that a crisis might arise in his country due to rivalry of local samurai leaders. He let his three sons band together, like a bundle of three arrows, to counter the threat. His reasoning was that one arrow can easily be broken. A bundle of three arrows cannot be broken.*
> *Yamamori & Mito (1993)*

SKI has, in a few recent events at Nuclear Power Plants (NPPs) in Sweden, found indications of weaknesses in the licensees' ability to handle weak signals, or diffuse symptoms of problems, and an inability to take proper actions in a reasonable time.

Problem Awareness

Most of the problems that arise in everyday operations are, on the whole, anticipated and therefore easy to recognize and manage. Operations as well as the rest of the organization (maintenance, technical and chemical analysis, etc.), try to keep up a constant state of unease and wariness in its ongoing search for deviations. In addition, clear-cut ways of reporting the discovered problem must be present for everyone to follow. The reason is simply that a problem that stays with whoever discovers it, is a problem that remains unknown.

The Swedish Nuclear Power Plants (NPPs) were good at identifying problems, including the more diffuse deviations, so the issue here was not a problem of perceiving weak signals. The problem was to translate the signal into something that could be seen as a safety issue with a potential to grow much larger.

Forum for Consultation

Different parts of the NPP organization have their own meetings and ways of discussing and analyzing occurring problems. There are also many decision forums on different levels of the organizations. One major meeting, where representatives from different parts of the organization are invited, is the daily planning meeting, which is held both during general operations and outages. This is an important part of the organizational structure that supports the immediate evaluation of concern.

The daily morning meetings served to state the unit's operational readiness and were generally considered efficient. One factor that affected the efficiency of this meeting was that it was considered a meeting of the operations department. Although other departments were present, the forum was not treated as an opportunity for representatives from other parts of the organization to put problems on the table and give a diffuse problem a chance to get consulted by the different competencies present. A contributing factor was that it was optional for representatives from the other parts of the organization to attend the meeting. Representatives from the independent safety department present at the meetings were, most of the time, just silent observers and did not comment on issues even when it was clear that some input could have been useful.

This meant that the daily planning meeting was a missed opportunity to get different opinions on potential safety-critical issues out in the open. Neither were decisions looked at from different viewpoints. Diffuse problems could therefore slip through the organizational structure of meetings.

Interestingly enough, this daily morning meeting was seen as an important meeting with the potential to catch vague problems. Yet what the NPPs thought was a cornerstone in an efficient structure for managing problems, was in fact a 'barrier' full of holes. The imagined efficiency of the meeting was not actually grounded in reality.

Strengthening the Forum

The daily meeting needed better routines in place to make the forum stronger and actually live up to its potential. The attendance of representatives from key departments of the organization was made obligatory. The agenda was updated with checkpoints where every representative had the possibility, and obligation, to report potential safety problems – diffuse symptoms that might need further analysis or action – so the forum could discuss and decide on a line of action. Decisions taken since the last meeting, regarding vague problems, were now expected to surface at the meeting. The focus clearly shifted from the operations view to a system view of the problems popping up in the meeting. It is emphasized that detected problems somewhere in the organization must be made known and should be given full attention in a forum where different competencies are present.

Of course, this daily planning meeting can lead to other investigative meetings where people from different parts of the organization are present and this type of meetings can also be held *ad hoc* at odd hours whenever it is necessary. But the daily planning meeting remains the forum where the current problems should be discussed to get the overall view of the state of the reactor unit.

Other Fora

Corresponding meetings at higher levels of the organization are as important as the one described above. These meetings must also have the same type of structure and process. The meetings' members should have different organizational background and perspectives and act as 'opponents' to decisions taken.

Any discovered unclear deviation, especially problems originating in parts of the organization other than operations, must not lead to operating with anomalies with unknown causes. Strange deviations need the attention of many parts of, and all levels of, the organization for fast action. Different organizational parts and levels that act as opponents to decisions taken or ongoing analysis, reduces the risks for staying on the wrong operating track and makes responses to weak signals and diffuse safety states and unforeseen events easier.

Yet just to have different forums is not sufficient. More important than that is *what* information is discussed and *how* it is communicated at meetings between *which* attendees.

A Bundle of Arrows

This text has only looked into one of several ways of keeping the organization more resilient; in this case keeping up the resistance to vague problems by having a clear structure for dealing with these problems early on, thereby preventing them from growing into big safety issues. My ending is like my beginning: When different competencies band together like a bundle of arrows the ability to counter threats and remain strong is improved and the organization and its defences will rebound from threats instead of being broken.

Chapter 11

Organisational Resilience and Industrial Risk

Nick McDonald

Introduction

Is resilience a useful term for describing how or whether an operational system functions effectively over an extended period of time? First, it is helpful to explore what resilience could mean and how the term could apply to operations that carry the threat of disastrous failure. Examining the normal way in which operational systems (and their host organisations) function uncovers a stark gap between formal requirements and what actually happens and points to the important role of professionalism in compensating for dysfunctional organisational systems. Quality and safety management does not easily ameliorate these complex problems. This, in turn, poses a problem for understanding how improvement and change comes about in organisational systems. A more fundamental question concerns how technology innovation processes in complex systems could address the problems of operational life. The notion of resilience could help shape our understanding of how to address these conundrums, but significant gaps limit our knowledge about the resilience of operational systems.

What is the Nature of Resilience?

If we can apply the term resilience to organisations and the operations they support, then it should be clear to what this term refers. There are a number of obvious ways to use the term, for example:

- To refer to a property of a system,
- A post-hoc judgement on a successful outcome – so far (the system has survived therefore it must be resilient),
- A metaphor that maps only loosely onto organisational or system characteristics.

Certainly, if the concept of resilience is to aspire to have some use in diagnosing systems, let alone predicting outcomes, it needs to be anchored in some clearly describable characteristics – characteristics which transcend those that may have an occasional association with organisational success on one occasion or another. Again, using the opposition 'resilience – brittleness' as a general characterisation of organisational functioning will not be helpful unless it is more than a general analogy or metaphor drawing comparison with the dynamic characteristics of physical objects.

If resilience is a system property, then it probably needs to be seen as an aspect of the relationship between a particular socio-technical system and the environment of that system. Resilience appears to convey the properties of being *adapted* to the requirements of the environment, or otherwise being able to manage the variability or challenging circumstances the environment throws up. An essential characteristic is to maintain stability and integrity of core processes, despite perturbation. The focus is on medium to long-term survival rather than short term adjustment *per se*. However the organisation's *capacity* to adapt and hence to survive becomes one of the central questions about resilience – because the stability of the environment cannot be taken for granted. Therefore, the notion is important of being able to read the environment appropriately and to be able to anticipate, plan and implement appropriate adjustments to address perceived future requirements.

Thus, resilience seems to reflect the tension between stability and change in organisational and operational systems, mediated by the notion of appropriate adaptation. But what is appropriate and to what environment? Within this frame, the notion of environment is potentially very broad, including both local disturbances (which cause, for example, the risk of operational failure) as well as the social and commercial factors (for example, the requirement to compete) that determine what has to be done to ensure survival of the organisation itself.

Commercial and operational risks interact, but the way in which they interact is not necessarily simple. For example, the notion of 'safety margins' is often invoked to imply a linear relationship between cost-cutting and the size of a theoretical 'safety margin'. Safety margins are hard (perhaps impossible) to define, particularly where operational processes are characterised by routine and regular unofficial action. There are also different ways of adapting and cutting costs. For example, lean, low cost approaches to operations are often based on a focused approach to managing processes, which, while having the objective of stripping out redundant or 'unproductive' activity, may also direct attention to the fundamental safety requirements of those processes. Thus, there may be no necessary link between the degree of apparent redundancy in operational systems and greater safety. 'Lean' operational systems may be safer or less safe than traditionally managed operations and the nature of their vulnerability to risk may be different. It is precisely because there are different ways of designing and managing an operation and different ways of cutting costs or otherwise responding to environmental pressures of one kind or another, that it is necessary to drive down the analysis to ask specifically and precisely what characteristics of operational systems could, in fact, bring benefits in terms of resilience.

This then suggests a provisional definition of resilience. Resilience represents the capacity (of an organisational system) to anticipate and manage risk effectively, through appropriate adaptation of its actions, systems and processes, so as to ensure that its core functions are carried out in a stable and effective relationship with the environment.

System Characteristics of Stability and Change

Adopting the notion of an organisational system brings to the foreground the functional characteristics of systems. Organisational *systems* (as open systems) comprise inputs, transformation processes and outputs. It is important to show how the *manner* in which an organisation deals with the physical, social or economic material it encounters in its operating environment, leads to outcomes that maintain a stable (or otherwise positive) relationship with that environment. The real interest in this is where environments are not inherently stable, and where this instability brings a risk of catastrophic failure to the system. Thus, maintaining stability requires the capacity to adjust.

This adjustment can occur at different levels, for example, an operational group, the organisational system, groups of organisations contributing to a single operational system, and the fundamental technologies of such an operational system produced by an industry group. All of these levels imply radically different timescales for both operation and adjustment, from short-term operational decisions to the development of a new technology over a number of years. In principle all of these levels of analysis are relevant to resilience and are interdependent, as if each wider layer of the onion sets the context for how the layer inside can operate. Nevertheless, despite this interdependence, each layer may require quite distinct explanatory mechanisms. There may also be tensions and incompatibilities between the mechanisms that operate at these different levels. In order to understand the nature of resilience it is probably necessary to develop an understanding of these different levels and of the relationships between them.

As Amalberti (Chapter 16) has argued, it seems clear that the relationship between system and environment (i.e., risk) will change according to the *stage of development of the technology*. The stage of development of the *organisation* is just as important (i.e., how the human role in managing the technology is itself managed and controlled). These two are closely related.

In the context of organisational systems, resilience would seem, on the one hand, to depend on increasing standardisation. The following are examples of such tendencies:

- Stronger co-ordination of processes by routinisation of procedures in operations and organisational systems;
- Increased reliability through the removal of variance due to individual skill, and ensuring the substitutability of different people, through standardised selection and training;
- Ensuring, through supervision, inspection, auditing, etc. that standardisation of the work-process does control the actual flow of work;

- Better standardisation of the outputs of the process is made possible through better monitoring, recording of those outputs;
- Automation of routine or complex functions.

On the other hand, resilience seems also to require a certain flexibility and capacity to adapt to circumstances. Common organisational forms of this are:

- Informal work-practices, which are often unofficial;
- Distributed decision systems with local autonomy;
- Flexible/agile manufacturing systems which can adjust to changing demand;
- Technologies that enable rather than constrain appropriate human action and modes of control;
- Organisational systems that can manage feedback, learning and improvement.

It may be therefore that resilience is bound up with being able to successfully resolve apparent contradictions, such as:

- Formal procedures – local autonomy of action;
- Centralisation – decentralisation of functions/knowledge/control;
- Maintaining system/organisation stability – capacity to change;
- Maintain quality of product/service – adjust product/service to demand or changing need;
- Use well-tested technologies – develop innovative technical systems.

In order to explore how some of these contradictions are managed in practice, we will use some examples from our own research to focus on the following issues: reconciling planning and flexibility in operational systems; the role of quality and safety management in fostering improvement; the problem of organisational change and the role of human operator requirements in industrial innovation.

It should be borne in mind, of course, that the successful reconciliation of these contradictions is not just a matter of internal consistency within the operation or organisation, but crucially concerns enabling particular ways of managing the interaction with the environment. This is a highly complex interaction. Take organisational theory as an example. Contingency theories, which posit an optimal fit between organisational form and environment, have been supplemented by recognition of the importance of choice of managerial strategy. This in turn has been modified by an acknowledgement of the power of large organisations to control and modify their environments, both through the dominance of the markets in which they operate and through their ability to create significance and meaning surrounding their areas of activity (for contrasting analyses see Thompson & McHugh, 1995, chapter 3; and Morgan, 1986, chapter 8). To adapt this line of argument to the notion of resilience, it seems clear that there is unlikely to be 'one best way' to achieve resilience, or even several 'best ways'. It is important however to understand how organisations, and those that manage them make choices, and what the consequences are of these choices. It is also important to recognise that achieving 'resilience' (whatever that means) is not just a matter of finding the right technical solution to an operational problem, but of constructing a better way of understanding the operational space. Has the notion of resilience the conceptual potential to sustain a re-engineering of our understanding of the way in which organisations and the people within them cope with danger, difficulty and threat?

Planning and Flexibility in Operational Systems

For the past number of years we have been exploring, through a series of European projects concerning aircraft maintenance, the notion of a symbiotic relationship between the functional characteristics of an organisational/operational system and the behaviour and activity of the actors within the system. While this is a conceptually simple idea, it requires the accumulation of evidence from a diversity of sources in order to substantiate it. These sources include, amongst others, studies of self-reported compliance with procedures in task performance, observations of task performance, analysis of documentation and information systems, cases studies of planning and quality systems, and surveys of climate and culture.[1]

Compliance with Procedures

The idea originated in a study of compliance with documentation and procedures. A survey of aircraft maintenance technicians in four different organisations established that in approximately one third of tasks the technician reported that he or she had not followed the procedure according to the maintenance manual. Most commonly the technicians reported that there were better, quicker, even safer ways of doing the task than following the manual to the letter. This is a paradoxical result. On the one hand, in a highly regulated, safety critical industry like aircraft maintenance it seems surprising that such a high proportion of technicians admit to what is essentially an illegal act, which violates the regulations. On the other hand, virtually no-one who knows the industry well has expressed any surprise at this result. Follow-up case studies demonstrated that it is indeed not very hard to find 'better' ways than following the manual exactly. It is also not hard to find examples of where an alternative method to the manual, which has been followed, is clearly not better or carries an undesirable risk. The manuals themselves are not an optimum guide to task performance – they have to fulfil other criteria, such as being comprehensive, fully accurate and up to date. In many organisations the documentation is not presented in such a way as to provide an adequate support to task performance. What this establishes is simply that judgement based on experience is important, along with following procedures, in determining how tasks are done (McDonald, 1999).

So far this is not very different from evidence from other domains concerning compliance with procedures. Indeed the work of Helmreich and his colleagues (e.g. Helmreich, 2001; Klampfer et al., 2001; Klinect et al., 2003; Helmreich & Sexton, 2004) in the development and deployment of LOSA (Line Operations Safety Audit) in a variety of airlines has demonstrated the normality and pervasiveness of error and

---

[1] The ADAMS, AMPOS, AITRAM and ADAMS2 projects were funded by the European Commission under the 4th and 5th Framework RTD Programmes.

procedural deviation in flight operations and in medicine. This is an important development for two reasons. First LOSA represents a system audit procedure for recording 'normal operational performance'. While Dekker (2005) might call this a contradiction in terms (because, for example, the act of observation transforms what is observed) LOSA represents a serious attempt to get nearer the observation of a normal situation by using elaborate mechanisms to protect flight crew from organisational jeopardy, preserve confidentiality, ensure remote analysis of aggregated data, *etc*. Thus for the first time there is a serious, if not perfect, window on an aspect of operational life, which, on the one hand, everyone knows about, but, on the other hand has been hidden from scrutiny and analysis. The second reason follows from this. The evidence from LOSA has been important in contributing to a new view of 'human error' in which error and deviations are seen to be routinely embedded in normal professional practice, and in which the practice of professionalism is concerned less with the avoidance of error *per se*, than with managing a situation in a way that ensures that the envelope of acceptable performance is commensurate with the context of potentially destabilising threats and that any 'non-standard' situation is efficiently recoverable. This is the Threat and Error Management model.

Planning and Supply

However, our studies of aircraft maintenance provide a wider context for examining professional behaviour. In some ways the flight deck or the operating theatre may provide a misleading model for contextual influence on behaviour precisely because the immediate context of that behaviour is so clearly physically delineated and apparently separated from the wider system of which it is part – the flight deck is the exemplar *par excellence* of this. On the contrary, aircraft maintenance operations are clearly intimately located in a wider technical and organisational context, which includes the hangar, airport, maintenance organisation and airline.

Aircraft maintenance is a complex activity. It can involve a maintenance department of an airline performing mainly 'own account' work or fully commercial independent maintenance organisations whose work is entirely third party. There are inherent contradictions to be resolved – for example, how to reconcile the requisite inventory of parts and tools to cover all foreseeable planned and unplanned work, while containing capital costs. A major check may involve thousands of tasks and take several weeks to complete.

For many organisations there appears to be an unresolved tension between effective planning and the requirement of flexibility to meet the normal variability of the operational environment. Some of the dimensions of this tension can be illustrated on some of our studies of aircraft maintenance organisations. The ADAMS project included a comparison of the planning systems of two organisations (Corrigan, 2002; McDonald, 1999). In one organisation, planning was done in a very traditional top-down hierarchical manner; in the second there were serious attempts to create a more responsive and flexible planning process. In the former the maintenance plan originates in the engineering department (following the requirements of the manufacturer's master maintenance plan). The plan moves down through the system through schedulers to the operational team on the hangar floor where it is apportioned out in the form of task cards to the technicians actually carrying out the work. Within this case study, nowhere is there an opportunity for feedback and adjustment of the plan itself in response to unexpected developments, problems or difficulties in the operational group. The second case concerns an organisation in the process of redesigning its planning system along explicit process lines, building in co-ordination mechanisms for the relevant personnel along the process so that plans could be validated and adjusted according to changed circumstances. Comments from those contributing to the case study indicated that while not all the problems were solved this was a big improvement on what had gone before.

It is when one examines what happens in the check process itself that one finds the consequences of problems of planning and also of the various supply chains that deliver tools, parts and personnel to the hangar floor. For those at the front-line (skilled technicians and front-line managers) problems of adequate supply of personnel, tools, parts, and time seem to be endemic, according to one study, whereas, ironically, problems of documentation and information are not seen to be a problem (presumably because these are not relied on for task performance). Management appear to spend a lot of time and effort 'fire-fighting' in a reactive manner, are not confident in managing their staff, and tend to engage in various protective stratagems (e.g. holding staff or tools) to protect the efficiency of their group but which also exacerbate instability elsewhere. There is also a certain ambiguity in the roles of technical staff – for example in relation to precise understanding of what is involved in supervising and signing off the work of others (only those appropriately qualified can certify that the work has been done according to the required procedures) (Ward et al., 2004).

Professionalism

In organisational analysis and diagnosis often the easy part is to identify the apparent imperfections, deficiencies and inconsistencies to which most organisations are subject. What is less obvious may be what keeps the system going at a high level of safety and reliability despite these endemic problems. In many systems, what is delivering operational resilience (flexibility to meet environmental demands adequately) is the 'professionalism' of front-line staff. In this context, professionalism perhaps refers to the ability to use one's knowledge and experience to construct and sustain an adequate response to varying, often unpredictable and occasionally testing demands from the operational environment. Weick (2001) has written well about this, using the term 'bricolage' to convey the improvisational characteristics that are often employed. However, it is important to keep in mind the notion that such behaviour is a function of both the environmental demands on task performance and the ways in which the organisational system supports and guides task accomplishment (or, relatively, fails to do so). The way in which this dynamic relationship between action, system and environment is represented in professional cultural values is illustrated in a meta-analysis of some studies of aircraft maintenance personnel.

Abstracting from a number of surveys of aircraft maintenance personnel in different organisations led to a generalisation that pervasive core professional values included the following characteristics: strong commitment to safety; recognising the importance of team-working and co-ordination; valuing the use of one's own judgement and not just following rules; being confident in one's own abilities to solve problems; having a low estimate of one's vulnerability to stress; and being reluctant to challenge the decisions of others. What does this tell us? The significance of these values suddenly becomes rather more salient when they are counterposed to some of the system characteristics of the maintenance organisations to which many of these personnel belong. As well as the problems of planning and supply outlined above, such organisations were not notable for their facility in solving problems (see below). Thus, these characteristic professional values seem, in many ways, to match the deficiencies found in such organisations. Essentially, and perhaps crudely, the professional values can be seen to reflect a philosophy that says: rely on your own resources, do not expect help or support and do not challenge, but above all get the job done and get it done safely. Seen in this light, professionalism compensates for organisational dysfunction, it provides resilience in a system that is otherwise too rigid. We have adopted the label 'well-intentioned people in dysfunctional organisations' to characterise this systematic pattern of organisational life.

Of course it does not always work out as intended. One of the prototypical anecdotal narratives from all around this industry goes like this. A serious incident happens in which an experienced technician is implicated – there was pressure to get the job done and the official rules and procedures were not followed. Management are very surprised that 'one of the best technicians' had committed, apparently intentionally, such a serious violation. It seems inexcusable, and incomprehensible, until one reflects on what it is that being 'one of the best technicians' actually involves. The role of such people is to use their knowledge and experience in the most judicious but effective way to get the job done. Questions are only asked about precisely what was done, following something going wrong. On other occasions informal, unofficial practices are everywhere practised, universally known about but absolutely deniable and hidden from official scrutiny.

Road Transport

There are many other examples of this syndrome. Some of our recent research concerns road goods delivery in urban environments (Corrigan et al., 2004).[2] This urban multi-drop road transport operation is a highly competitive business that operates over a small radius. The multiple stops are arranged according to very tight schedules, which are dictated by the customer's requirements, rather than the driver's optimal logistics. Access is a major problem, with a lack of loading bays in the city centre, the absence of design requirements for ensuring good access, and, often, poor conditions at customer premises. The traffic system in which these drivers operate is congested, with complex one-way systems and delivery time restrictions. Clamping, fines and penalties are routine hazards. Coping with the multi-drop system reportedly cannot be done without traffic violations, particularly parking violations and speeding. Interestingly, in many companies, while the driver is liable for speeding penalties the company pays for parking violations. The constant time pressure leads to skipping breaks. Neither the working-time nor the driving-time regulations effectively control long working hours. Of course, the driver has to compensate for every problem, whether it relates to wrong orders, waiting for the customer, or delays due to congestion.

A high proportion of daily tasks are rated as risky by drivers. Risky situations are rated as high risk to self and others, high frequency, and high stress, and drivers have no control over them. For example, unloading while double parking in street involves risks from other traffic, loading heavy goods, risks to pedestrians, and a rush (to avoid vehicle clampers). Professionalism is managing these risks to ensure as far as possible a good outcome. Again, arguably it is this characteristic of the operating core – the professionalism of drivers – which gives properties of resilience to a rigid system that is not optimally designed to support this operation.

Organisational Differences

This discussion has focused on what may be, to a greater or lesser extent, common characteristics of humans in organisational systems. It

---

[2]   Safety and Efficiency in Urban Multi-drop Delivery is part of the Centre for Transportation Research and Innovation for People (TRIP) at Trinity College Dublin, and is funded by the Higher Education Authority.

is clear however that organisations differ quite markedly. Our research in aircraft maintenance organisations explored some of these differences and, again taking a global view, has given rise to certain intuitive generalisations. For example, certain organisations devote more effort and resources than others to the planning process. It sometimes seems as if this commitment to planning comes with an expectation that the plans will work and a reluctance to acknowledge when things go wrong. On the other hand, those organisations who are less committed to planning, have to learn to be flexible in managing, largely reactively, the normal variation of operational demands. It is tempting to suggest that some of these differences may be influenced by regional organisational cultures – but this is the subject of ongoing research and it is too early to make confident generalisations.

Looking at the wider literature, a contrasting analysis of what may be highly resilient systems has been given by the High Reliability Organisations group. This emphasises, for example, distributed decision-making within a strong common culture of understanding the operational system and mutual roles within it. Descriptions of this work (for example Weick, 2001) emphasise the attributes of organisations and their people, which characterise the positive values of high reliability. But some attributes of organisations may have positive or negative connotations depending of the context or depending on the focus of the investigation. Thus, the notion of informal practices based on strong mutual understanding can be seen in the context of Weick's analysis to be one of the characteristics of high reliability in organisations. In another context (for example Rasmussen, 1997; Reason, 1997), which emphasises the importance of procedures and standardisation such practices can be seen to be the basis of systematic violations. A large part of this contrast may be differences in theoretical orientation with respect to the nature of performance, cognition, or intention. However, part of the contrast may also relate to real differences in organisational forms. It is impossible to know how much of this difference is due to fundamental organisational differences and how much is due simply to looking at organisations from a particular point of view.

At the operational level, therefore, resilience may be a function of the way in which organisations approach and manage the contradictory requirements of, on the one hand, good proceduralisation and good planning, and on the other hand, appropriate flexibility to meet the real demands of the operation as they present on any particular day. Different organisations, different industries will have different ways of managing these contradictions. However from the point of view of our own research with aircraft maintenance organisations, one problem is that the 'double standard' of work between formally prescribed and unofficial ways of working is hidden. While everyone knows it is there, the specifics are not transparent or open to scrutiny or any kind of validation. It is clear that the unofficial way of working both contains better (quicker, easier, sometimes safer) ways of working as well as worse ways. Sometimes these informal practices serve to cover over and hide systematic deficiencies in the organisation's processes. Thus, a major issue is the transparency of informal practices – are they open to validation or improvement? It is important that transparency is not just to colleagues but also to the organisation itself. Therefore we need to look at the capacity of organisations to deal with informal and flexible ways of working. One aspect of the necessary change concerns the capacity to change rules and procedures to deal with specific circumstances. There appear to be few examples of this. Hale et al. (Chapter 18) cites an example in a chemical company and Bourrier (1998) describes a process by which the maintenance procedures in one nuclear power plant (in contrast to some others) are revised to meet particular circumstances.

However, the issue is not just one of procedural revision, but of ensuring that all the resource requirements are met for conducting the operation well. Indeed, the adequacy of documentation and information systems may, on occasion, be seen to be the least critical resource. Thus, the issue in not just one of technical revision of documentation but of the capacity to adapt the organisation's systems (including those of planning and supply) to optimally meet the requirements of the operation. Thus the 'resilience' of the operational level needs to be seen in the context of some oversight, together with the possibility of modification, change or development at a wider organisational level. Although this could be done in a variety of different ways it makes sense to ask: what is the role of the quality and safety functions in an organisation in maintaining or fostering resilience?

The Role of Quality and Safety in Achieving Resilience

The argument is, therefore, that to be resilient, the operational system has to be susceptible to change and improvement – but what is the source or origin of that change? From some points of view quality and safety management are about maintaining stability – assuring that a constant standard of work or output (process and product) is maintained. From another point of view these functions are about constant improvement and this implies change. Putting both of these together achieves something close to the provisional definition of resilience suggested earlier – achieving sufficient adaptation in order to maintain the integrity and stability of core functions. How does this actually work in practice? Again some of our studies of aircraft maintenance organisations can help point to some of the dimensions that may underlie either achieving, or not achieving, resilience. The point is not to hold up a model of perfect resilience (perhaps this does not exist), but to use evidence from real experience to try to understand what resilience as a property of an organisational system might involve.

The Role of Quality

The traditional way of ensuring quality in the aircraft maintenance industry has been through direct inspection, and in several countries within the European Joint Aviation Authorities system, and in the United States under the FAA, quality control through independent direct inspection of work done is still the norm. In some companies the quality inspectors or auditors are regarded as policemen, with a quasi-disciplinary function, while in others the role of auditing or inspection is seen to include a strongly supportive advisory role. However, the essence of the current European regulations is a quality assurance function – the signing off for work done (by oneself or others) by appropriately qualified staff. Fundamentally this is the bottom line of a hierarchical system based on the philosophy of self-regulation. In this, the National Authorities, acting under the collective authority of the JAA (and now of the European Aviation Safety Authority – EASA), approve an operator (maintenance organisation) on the basis that they have their internal management systems in place (responsible personnel, documented procedures, etc.). It is this system that is audited by the authorities on a periodic basis. How far does this system of regulation contain mechanisms that can foster change and improvement?

None of the quality systems we have examined contain a systematic way of monitoring or auditing actual operational practice. Thus, the 'double standard' between the official and the actual way of doing things becomes almost institutionalised. There is a paper chain of documentation that originates in the manufacturer's master maintenance plan, which moves through the maintenance organisation (engineering, planning, scheduling, operations, quality) and then is audited by the national authority. This document chain of accountability only partially intersects with what actually happens on the hangar floor or airport ramp, because there is no effective mechanism for making transparent actual operational practice, let alone a system for actually reducing the distance between the formal requirements and actual practice.

It is a requirement for maintenance organisations to provide an opportunity to give feedback on quality problems. While this may be a good idea in theory, our evidence suggests that it is not working in practice. Not all of the organisations we studied had a fully operational quality feedback system in place. However, even where there was a system that gathered thousands of reports per year, there was little evidence that feedback actually influenced the operational processes which gave rise to the reported problems.

Improvement

In order to try to address this problem, a specific project was undertaken in order to examine the improvement process. This project was called AMPOS – Aircraft Maintenance Procedure Optimisation System. This was an action research project based on a simple idea. A generic methodology moved individual cases (suggestions for improvement in procedure or process) through the improvement cycle (involving both maintenance organisation and manufacturer). The methodology was instantiated in a software system and implemented by a team of people who managed these cases at different stages in both organisations. A set of cases were gathered and processed through the system.

Two very broad conclusions came out of this effort. First, that problems involving the human side of a socio-technical system are often complex and, even if conceptually tractable, are organisationally difficult to solve. Although it is nonsensical to separate out problems into 'technical' and 'human' categories, as a very rough heuristic it makes sense to think of a continuum of problems along a dimension from those that are primarily technical in nature to those that are primarily human. At the latter end, the potential solutions tend to be more multifaceted (for example involving procedure, operational process, training, etc.) and require significant change of the operational system and the supporting organisational processes. Furthermore, it was not always clear exactly how these problems could (or should) be solved or ameliorated at the design end – and it became clear that this was a much more long-term process.

The second conclusion was that each succeeding stage of the improvement cycle was more difficult than the last. Thus, while it was relatively easy to elicit suggestions for improvement, it was rather more difficult to facilitate a thorough analysis that supported a convincing set of recommendations for change. Such recommendations have to be feasible and practicable to implement, which creates yet more profound problems, and even when such recommendations are accepted by those responsible for implementing them, it is even more difficult to get these recommendations implemented, particularly those that were more complex and challenging. As for evaluating the effectiveness of the implementation, this turned out to be beyond the feasible timescale of the project (McDonald et al., 2001).

Response to Incidents

The messages here are much more general than this particular project, however, and similar fundamental problems can be seen in relation to the manner in which organisations manage their response to serious safety incidents. An analysis of a series of incidents of the same technical/operational nature in one airline posed the questions: what was done after each incident in order to prevent such an incident happening again? Why was this not sufficient to prevent the following incident? One interpretation of the changing pattern of response to each succeeding incident was that there seemed to be a gradual shift from a set of recommendations that were technically adequate to recommendations that were both technically and operationally adequate – they actually worked in practice. This case study furthermore demonstrated that it is extremely difficult to get everything right in practice, as unforeseen aspects often come around in unexpected ways to create new possibilities for incidents. In this case, each succeeding incident appeared to have challenged the mindset of the team responsible for developing an appropriate response, disconfirming initial assumptions of what was necessary to solve the problem, and in this case, eventually involving the operational personnel directly in finding an adequate solution (McDonald, 1999).

This particular case concerned an airline with an exemplary commitment to safety and well-developed safety infrastructure (a dedicated professional investigation team, for example). For other organisations who are not as far down the road in developing their safety system, the cycle of organisational activity is a rather shorter one. Some organisations have not been able to satisfactorily reconcile the organisational requirement to establish liability and to discipline those at fault with the idea that prevention should transcend issues of blame and liability. Others have adopted an uneasy compromise, with an official 'no-blame' policy being associated with, for example, the routine suspension of those involved in incidents, normally followed by retraining. A lot of organisational effort goes into these different ways of managing incidents. The more apparently serious or frequent the incident the more necessary it is for the organisation to be seen to be doing something serious to address the problem. Such organisational effort rarely seems to tackle the underlying problems giving rise to the incidents. On the contrary such 'cycles of stability' appear to be more about reassuring the various stakeholders (including the national authorities) that something commensurate is being done, than with really getting to the bottom of the problem.

The issue for resilience is this. We do not seem to have a strong, empirically based model of how organisations respond, effectively or otherwise, to serious challenges to their operational integrity, such as are posed by serious incidents. The evidence we have gathered seems to suggest that 'organisational cycles of stability' (in which much organisational effort goes, effectively, into maintaining the status quo) may be more the norm than the exception. Even in apparently the best organisations it can take several serious incidents before an operationally effective solution is devised and implemented. On a wider scale, there is a disturbing lack of evidence in the public domain about the implementation of recommendations following major public enquiries into large-scale system failures. One exception to this is the register of recommendations arising out of major railway accident enquiries maintained by the UK Health and Safety Executive. In this case, when and how and if such recommendations are implemented are publicly recorded. Thus, by and large, we really do not know how organisations respond to major crises and disasters.

The Management of Risk

The concept of resilience would seem to require both the capacity to anticipate and manage risks before they become serious threats to the operation, as well as being able to survive situations in which the operation is compromised, such survival being due to the adequacy of the organisation's response to that challenge. We know far too little about the organisational, institutional and political processes that would flesh out such a concept of resilience. More examples and case studies are needed of the processes involved in proactive risk management and safety improvement. Such examples should encompass not only the identification of risk and prioritising different sources of risk, but also the assignment of responsibility for actively managing the risk, the implementation and evaluation of an action plan and monitoring the reduction of risk to an acceptable level.

In a local example of such an initiative, Griffiths & Stewart (2005) have developed a very impressive redesign of an airline rostering system to reduce the risks associated with fatigue and the management of rest and sleep. This case was based on a clear identification and measurement of the extent of the risk, leading to the implementation of a redesigned roster system together with the continued monitoring of a number of indices of risk to ensure that the sources of risk were effectively controlled.

A systemic approach to quality and safety can help address these problems. Analysing the organisational processes of quality and safety as a system which receives inputs, transforms these and produces outputs can help focus attention on what such processes deliver to the organisation in terms of improvement. Following this logic, we have produced generic maps of two quality and safety processes, based on the actual processes of several aircraft maintenance organisations (Pérezgonzález et al., 2004). Further to this, Pérezgonzález (2004) has modelled the contribution of a wide range of organisational processes to the management of safety.

However, these examples do not really encompass the complex processes of organisational change that appear to be necessary to create what one might call an adaptive resilient organisation.

The Problem of Organisational Change

Even apparently simple problems of human and organisational performance require complex solutions. As we have seen, operational environments are not always designed to optimally support the operator's role. The key to understanding the possibilities of change concerns this relationship between system and action – how the organisation's systems and processes are understood as constraining or allowing certain courses of action in order to meet the contingencies of a particular situation. However, even if we understand *what* may need to change, *how* to change poses a question of a different order of complexity. Even if it is accepted that there is a problem, it cannot be taken for granted that change will occur. The problem of organisational systems which are apparently highly resistant to change despite major failure have been particularly strongly brought to public focus in the establishment of enquiries into the Hatfield train accident in the UK and the Columbia shuttle disaster in the US.

From our evidence, for many organisations, inability to change may be the norm. We have described 'cycles of stability' in quality and safety, where much organisational effort is expended but little fundamental change is achieved. Professional and organisational culture, by many, if not most, definitions of culture, reinforces stasis. This active maintenance of stability is conveyed in the popular conception that culture is 'the way we do things around here'. The complexity of operational systems and the coupling and linkages between different system elements are also factors that would seem to militate against a view that change should be easy or straightforward. From the point of view of resilience, that core of stability in the operational system and its supporting processes is undoubtedly an essential characteristic, enabling it to absorb the various stresses and demands that the environment throws up, without distorting its essential mission and operational integrity. Thus the argument is not that change is necessarily beneficial, *per se*, rather it is the appropriateness of change that is the issue. Resilience would seem to demand the possibility of adaptation and change to improve the quality and reliability of a system's interaction with the environment and to meet a variety of challenges: to face new threats or demands from that environment, to incorporate new technologies that change the nature of the operating process, to rise to competitive pressures that constantly alter the benchmark of commercially successful performance, or to respond to changing public expectations of what is an appropriate level of risk to accompany a particular social gain.

The overwhelming weight of evidence we have accumulated in the aircraft maintenance and other industries is that change is necessary, but also that change is inherently difficult. The slow deliberate process of change to create more resilient systems is evident from Hutter's (2001) analysis of the implementation of safety regulation through effective risk management in the British Railways system, following the major accidents at King's Cross and Clapham Junction, and just prior to privatisation. Hutter identifies three phases of corporate responsiveness to safety regulation. In the Design and Establishment phase, risk management systems, procedures and rules are set up, committees established and specialists appointed. The main people involved at this stage are senior management and relevant specialists. In the Operational phase, these systems, procedures and rules are operationalised or implemented. Committees meet, audits are undertaken and rules are enforced. Those involved are primarily management at all levels and worker or community representatives. In the third phase – Normalisation – compliance with risk management rules and procedures is part of normal everyday life. Everyone in the corporation is involved, with both corporate understanding and individual awareness of risk (Hutter, 2001, p. 302). What is well described in Hutter's book is the process of moving from the Design and Establishment Phase to the Operational Phase. However, the critical transition from the operational phase to the normalisation phase is not well understood, and in relation to the British railways, appears to have been made more difficult by the privatisation of a previously nationally owned industry. Thus there seems to be little evidence about the way in which organisations have made a transition towards a fully normalised self-regulatory risk management regime. This poses the central problem in our analysis of change in resilience – how to generate effective change to increase resilience in the operating core?

There are few empirically based models that describe the dimensions and dynamics of the organisational change process. One such model was developed by Pettigrew & Whipp (1991) from an analysis of organisational change in a variety of industries in the UK in the 1980s. One of the compelling features of this model is that it puts the processes of organisational change firmly in the context of the competitive pressures in the commercial environment that drive such change. This perspective leads to a powerful analysis that seeks to demonstrate how the various dimensions of the change process contribute to meeting that environmental challenge. It is also a model of quite radical change, which is also appropriate to our analysis of resilience. Indeed it can be seen as a prototypical study of commercial resilience. The five dimensions of change which are highlighted in this analysis are environmental assessment, linking strategic and operational change, leadership, the role of human resources as assets or liabilities and coherence. Interestingly, the first of these, environmental assessment, is perhaps the least satisfactory. Pettigrew and Whipp emphasise the subjective, intuitive way in which senior management assess their commercial environment, which contrasts with the aspirations to rationality of a risk management approach, as outlined above. What is particularly interesting is the analysis of how that strategic imperative, which arises out of environmental assessment, is linked to the particular form of operational change that will deliver competitive advantage. Successfully achieving such operational change requires both intensive and multifaceted leadership support at all levels, and a programme of human resources development to ensure that the people in the equation become assets, rather than liabilities, to the change process. The final dimension – coherence – recognises that successful organisational change is complex and requires considerable effort to ensure that the different strands and levels of the process actually do work together to support the same direction of change.

This model poses a core challenge of theories of resilience to demonstrate not only how particular operational forms deliver outputs that have characteristics of resilience, but also what organisational requirements are necessary to implement these forms where they do not already exist.

Change in Technology

Arguably, in the long term, a more fundamental issue than that of organisational change is the issue of increasing resilience through technological innovation. To take air transport as an example, the impressive record of the aviation industry in dramatically improving its major accident rate per mile travelled over the last fifty years or so is commonly attributed to the advances in technological reliability of aircraft systems. Nevertheless, the human (pilot, air traffic controller) continues to make a predominant contribution to aviation accidents. Therefore it is logical to argue that the next generation of aviation technologies requires a step change in design for human use if a significant reduction is to be made in the persistence of human action as a contributor to accidents. The importance of this is emphasised by the pace of technological change, where the increasing potential for the integration of different subsystems (flight deck, air traffic control, airport ground control, for example) means that the roles of the human operator can potentially be radically re-engineered in a relatively short time. In such complex systems it is the human operator who plays the role of the critical interface between different subsystems. To achieve this, it is necessary to regard the human operator as a core system component (at all stages of the processes that make up the system) and whose role and functions need to be addressed from the very start of the design process.

Achieving a 'system integration' model of human factors, however, is difficult in practice as well as in theory. In many systems there is a considerable distance – professional, cultural, geographic – between design and operation. To take one example, those who write or are responsible for revising aircraft maintenance procedures are often surprised when they hear of routine non-compliance with such procedures. Such procedures are written to be followed and the idea that, from the technician's point of view, there might be 'better' ways of doing the task than following the procedure to the letter often seems bizarre and irresponsible. Nevertheless, this behaviour makes sense to the technician. Maintenance manuals are designed to be comprehensive, accurate, up to date, and sufficiently detailed to allow an inexperienced mechanic to complete a task. This design requirement does not include being a user-friendly task support for the average technician. Designing information and documentation systems which more closely meet the way in which the task is actually performed as part of a maintenance process is a considerable challenge. The routine errors and violations uncovered by the LOSA methodology likewise pose a similar challenge to aircraft flight system designers. The same principle is true for those who design the road transport and goods handling infrastructure that is used by urban multi-drop drivers. The examples can be generalised to virtually any context of human activity within large complex technological systems. The question is: How to bring operator requirements to the forefront of the design process?

Currently we have been unable to find any convincing models of how user or operator needs can be constituted and put into play as a driver of innovation in large socio-technical systems. This seems to be a considerable gap in understanding how to develop resilience in new generation systems. This is therefore the topic of two current European projects, which address, amongst other things, the role of human factors in the aviation system.[3] The basic idea is as follows. It is important to capture knowledge about the operator in the system, not only about those at the so-called 'sharp end' of the operation, but all along the operational process. Much of the relevant knowledge is tacit knowledge about how the system operates and how people act within it. This is knowledge that is rarely made explicit or written down, but which is commonly known. This knowledge has to be transformed to be of relevance as a source of potential design ideas. The model for this knowledge transformation process is the knowledge transformation project teams, described by Nonaka (2002), whose task is to develop new product ideas. Such teams have to have the requisite variety and redundancy of real operational expertise to ensure that good ideas can be generated and thoroughly explored. Knowledge transformation is not sufficient on its own, however. In a large distributed multi-organisation system, like aviation, such knowledge has to be shared between organisations. The model for this process is the theory of industrial innovation pioneered by Porter (1998) and Best (2001), who see innovation as coming from the interaction of clusters of

---

[3]   Technologies and Techniques for New Maintenance Concepts (TATEM) and Human Integration into the Life-cycle of Aviation Systems (HILAS) are funded by the European commission under the 6th Framework programme.

organisations – in this case design and manufacturing organisations, operators and research and development organisations. Within our adaptation of this model, human factors research and development (both basic and applied developmental research) is fully incorporated into the innovation cycle, drawing its fundamental research topics from problems of innovation and change in system, process and technology. It is an ecological model of human factors research, which is grounded in operational reality, integrated in the systems, processes and technologies that structure the operation, and is interdisciplinary in setting the human requirements at the centre of a systems engineering perspective. The broad objective of this research is to model a system of innovation that can lead to the design of more resilient systems precisely because they are designed to meet operationally valid requirements derived from the actual practice of users. It is too early to see how far this will in fact be possible.

Conclusions – the Focus on Resilience

Resilience has been defined here in terms of a productive tension between stability and change. The basic stability and integrity of the system is an important dimension, as is the capacity to absorb major disturbances from the operating environment and to recover from failure. The notion of adaptation to the requirements of the operational environment implies the capacity to adapt and change in order to survive in a changing environment. The difficulty of understanding processes of adaptation and change is a recurring theme in this chapter.

The notion of resilience has to work on at least three levels – the operation (the individuals, group or team who work through the task and operational processes, with the relevant technology to produce the required result or output); the organisation (which incorporates, organises, co-ordinates, resources and in other ways supports the operations which produce the outputs that fulfil the organisations' mission); and the industrial system (which designs and produces the technologies that make the operation possible). A truly resilient system should absorb, adapt, adjust and survive at all three levels. Requirements are different at each of these levels and in some ways may be contradictory. Thus, for example, systems that rely on informal flexibility at an operational level are not always transparent at an organisational level. The independence of quality and safety systems from operational and commercial influence has to be reconciled with the need for quality and safety functions to be actively engaged with improving the operation and its processes. The contradiction has to be resolved between the organisational imperative to change to adapt to new circumstances or new events, and the sheer organisational effort and difficulty of successfully engineering such change. Again, such change should not in turn disrupt the core stability of operational processes, which are the central requirement of resilience. There are also many barriers to sharing, at an industry level, critical knowledge about how an operation really happens. Putting such knowledge to work in improving the next generation of systems poses further, as yet unanswered, questions.

However, resilience is not just about being able to change (on the one hand) or maintaining stability (on the other). It is critically about the appropriateness of stability or change to the requirements of the environment or, more accurately, about the planning, enabling or accommodating of change to meet the requirements of the future environment (as anticipated and construed) in which the system operates. This poses two challenges – both conceptual and practical. How to conceptualise and formulate hypotheses about the relationships between operational systems and the demands of their environments? How to test these hypotheses over the appropriate timeframe with appropriate longitudinal design?

If resilience is to be a useful theoretical concept it has to generate the research that will identify the particular characteristics of the socio-technical system that do in fact give it resilience. This chapter has tried to flesh out some of the relevant dimensions, but at crucial points there is a lack of sound empirical evidence that is grounded in operational reality, is systemic (located in its technical and organisational context), dynamic (i.e., concerns stability and change over time) and ecological (i.e., concerns systems in their environment). Unless this evidence gap is addressed, the concept of organisational resilience is in danger of remaining either a post-hoc ascription of success, or a loose analogy with the domain of the mechanical properties of physical objects under stress, which allows certain insights but falls short of a coherent explanation.

An Evil Chain Mechanism Leading to Failures

Yushi Fujita

Setting a higher goal for long-term system performance requires a variety of improvements such as improving yield, reducing operation time, and increasing reliability, which in turn leads to a variety of changes such as modifying the artifact (i.e., technical system), changing operating methods (either formally or informally), and relaxing safety criteria. These changes may implant potential hazards such as a more complicated artifact (i.e., technical system), more difficult operation, and smaller safety margins, which further may result in decreased reliability, decreased safety, and decreased long-term performance. The end results may be long interruptions or even the termination of business. It is important to understand that adaptive and proactive human behaviors are often acting as risk-taking compensation for the potential hazards behind the scenes, even though they are often admired as professional jobs. A resilient system can cut this evil chain by detecting and eliminating potentially hazardous side effects caused by changes.

Chapter 12

Safety Management in Airlines

Arthur Dijkstra

Introduction

Aviation has gone through a change and development from when 'the aircraft was made of wood and the men of steel' to a situation where 'the aircraft is made of plastic and operated by computers'. This refers to the many sides of change in the domain of aviation (which, of course also applies to other industries). Aviation has grown into a highly complex and dynamic industry where the margins for profitability are small. Airline companies are maximising production on minimal costs while trying to avoid accidents. Competition is fierce and cost cutting is the focus for economic survival. The environment to which an airline has to adapt is constantly changing. Airlines start network organisations and specialise in specific markets; rules and regulations are changing due to multilateral agreements where national laws are replaced by international regulations; and the technological change of the recent years has influenced aircraft operations and airline management strategies considerably.

In the early years, the men of steel had to deal with unreliable technical systems on board their aircraft. Accidents in those days were attributed to technical failure. This was a wide accepted cause since technical reliability was not very high. Over the years systems became more reliable, accurate and complex. Slowly accident causation explanations shifted from technology to the human operator. The prevailing view was that if machines performed as designed it must have been the human who failed. The conclusion of 'human error' as cause was accepted for some time, but now, increasingly, human error is the initiation of investigation and not the endpoint. Organisational as well as technological aspects have come into focus as contributing factors for accidents. This was the result of the recognition that human action does not occur in a vacuum but in a context of organisational and technological factors.

This is the background in which airline safety management has contributed to a safety level of one accident in about a million flights. Other main parties that contributed to safety improvement are the airline manufactures, Air Traffic Control services and regulators, but they are not discussed here any further. The increase in safety level has slowly declined and asymptotically approached its current apparent limit. I will discuss the developments in safety management, supported by interviews with accountable managers and safety practitioners. Some problems will be indicated and a wish list on how to move towards resilience engineering will be proposed.

How Safe is Flying?

To illustrate the safety of air travel the Boeing website gives the following data (Boeing 2005).

> In 2000, the world's commercial jet airlines carried approximately 1.09 *billion* people on 18 *million* flights, (Figure 12.1) while suffering only 20 fatal accidents.

> In the United States, it's 22 times safer flying in a commercial jet than travelling by car, according to a 1993-95 study by the U.S. National Safety Council. The study compares accident fatalities per million passenger-miles travelled. The number of U.S. highway deaths in a typical six-month period (about 21,000) roughly equals all commercial jet fatalities worldwide since the dawn of jet aviation four decades ago. In fact, fewer people have died in commercial airplane accidents in America over the past 60 years than are killed in U.S. auto accidents in a typical three-month period. For the year 2000, 41,800 people died in traffic accidents in the U.S. while 878 died in commercial airplane accidents.

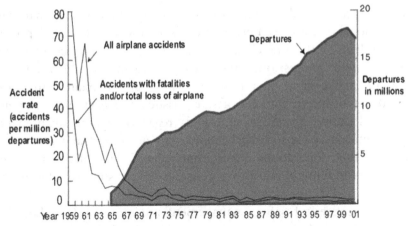

Figure 12.1: Accident rate history. Source: Boeing (2005)

A comparison between transport modes is shown in Figure 12.2.

Current Practices in Safety Management

ICAO (International Civil Aviation Organisation), IATA (International Aviation Transport Association) and the FAA (Federal Aviation Administration) in the United States and the JAA (Joint Aviation Authorities) are the legal bodies that regulated safety management in airlines.

An important issue in safety management is the place in the organisation of the safety department. This place influences the 'span of control' in the organisation by the number of department boundaries that have to be crossed horizontally or vertically. Regulations allow two ways of implementing a safety office into an airline organisation. One way of organisation is that the flight safety department is directly connected to the board of directors and directly reports to the COO (Chief Operating Officer) and consequently the department has a companywide scope of safety issues. The other is that each main organisational branch has its own safety department. The different safety departments then have to coordinate safety issues that pass organisational boundaries. Directors of the safety departments report to the president of the branch and always have a bypass access to the COO. Safety discussions that cross organisational boundaries horizontally, like flight operations talking to aircraft maintenance, can show different interpretations of safety issues. For example, an engine that failed during ground operation is regarded by the maintenance department as a pure technical failure with no impact at flight safety, while the flight operations department regards this failure as a flight safety related failure because they consider the question of what would have happened if the aircraft had just got airborne.

The practices of safety management are more or less the same across airlines. Airline safety management systems have to comply with FAA, JAA and IATA regulations, which are audited by audit programs. Leverage points for change in safety practices can thus be created by the changes in the regulations.

By analysing the practices at the safety department and the way information is used in the organisation to deal with risk and safety matters, the underlying theoretical concepts and models of risk and safety that these practices represent can be extracted. These models are not made explicit in the FAA, JAA, ICAO or IATA documentation nor in the safety department. People working at the safety department, including management, have no clear picture of why they do what they do. Practitioners and regulations do not refer to any underlying theoretical foundation for the practices. Comments on this observation are that it has always been done more or less like this and there are no alternative safety management practices, just some variations on common themes. Progress in safety and to break the current safety level cannot be expected to be initiated by the airlines themselves. They lack the resources, like time and knowledge for innovations, so help is needed from, e.g., the field of Resilience Engineering to supply alternatives to current practices. Since airlines follow the regulator requirements, regulations that include resilience engineering concepts might support progress on safety.

Data Sources that Shape the Safety Practices

Work in a safety department consists of data collection and organisation, composing management reports, conducting investigations into accidents or incidents and feeding line management with safety related information. The different sources are listed and later an example will be discussed on how information is presented to management. Various sources of data are used and combined in the attempt to create a picture about current risks and safety threats.

The confidentiality of operators, such as pilots, has consequences for data collection. Protocols for data collection and usage have been agreed upon with the pilots' unions. Information about pilots' names may never get known outside the safety department. In addition, data may never be used against pilots unless there is gross negligence or wilful misconduct (which is of course hard to define). Such pilot protection positively supports the willingness to report, the fear of blame is minimised and more reports about safety matters will likely be submitted to the safety department.

Air Safety Reports. A critical issue in data collection from reports is what events are reported and how this is done. The commander of an aircraft may always use his own judgement on what events to report and they are supported by an operating manual, which states what events must be reported. Airlines have different sorts of incidents that are defined as occurrences associated with the operation of an aircraft, which affects or could affect the safety of operation. A partial list defining incidents is shown in Figure 12.2.

```
Air Traffic incident
        A serious incident involving Air traffic Control
Technical incident
        Damage or serious cracks, corrosion etc found during maintenance or inspections
        Fires and false fire warnings
        Contained engine problems
        Failure of onboard systems requiring non-normal operating procedures
Ground incident
        Collision with cars, terrain etc.
Operational incident
        Not being a technical or air traffic incident
        Incapacitation of a crewmember
        Rejected take-off
        Exceeding operational limitation e.g. hard landing, over speed, low speed
        A significant error in fuelling or load sheet
        Critically low in-flight fuel quantity
        A significant deviation from track of flight level
        Landing at airport other than destination except for reasons of weather
```

Figure 12.2: Examples of incident description

Some occurrences are also triggered by other data collection methods (e.g., automatic aircraft data registration) which will be mentioned later. When an event is not triggered by another data collection method, the issue of reporting culture becomes important. Reporting is an aspect of a safety culture (Reason, 1997). In a good safety culture, reporting of safety threats is welcomed and supported. Conversely, relevant safety information may never get into the safety department when pilots are not willing to report or do not see a use in reporting.

Currently, reports are paper forms filled out by pilots. In the future, on-board computers may be used for filling out the report. The boxes to tick and comments to make should give a representation of the event. The contextual description of the event is done with aspects such as, time, altitude, speed, weather, aircraft configuration etc. The event description itself is a block of free text and the information to include is up to the writer.

Some pilot reports are very short and minimal information is given by the pilots. This may give the impression to the reader that the report was written just to comply with the rules and that it was not regarded as a serious event by the pilots. Also professional pride in their work may make pilots reluctant to submit long reports explaining their own (perceived) imperfect work. Maybe some pilots have some fear of negative judgement about their performance. Incidents are assessed at the safety office and pilots may be invited to see a replay of the flight data if the safety officer decides so. For this process a protocol is established between the pilot unions and the airlines. This process will not be the same across all airlines since this is a delicate issue for the unions because of the fear of pilot prosecution. Recent developments fuel this fear. In Japan, a captain was prosecuted when a flight attended got wounded and finally died after her flight had gone through heavy turbulence. In the Netherlands, two air traffic controllers were prosecuted when an aircraft taking off almost collided with an aircraft crossing the same runway. It takes a long time before these fears fade away. There is a saying among pilots that they do not want to face the 'green table'. This is a ghost image of being held responsible for some mishap on the basis of hindsight. There is a feeling of unfairness that that pilots have to take their decisions sometimes in split seconds and that the investigation committee can take months and have knowledge of the outcome.

Air safety reports are kept in specially designed database programs. Each event is assigned a risk level according to the IATA risk matrix (Figure 12.3) or an own matrix developed by the airline.

Likelihood or Probability that an incident / accident or damage occurs	Severity / Scope of Damage				
	Insignificant No or minor injury or negligible damage	**Minor** Minor injury or minor property damage	**Moderate** Serious but non-permanent injuries or significant property damage	**Critical** Permanent disability or occupational illness or major property damage	**Catastrophic** May cause death or loss of property
Often	Medium	High	Substantial	Substantial	Substantial
Occasionally	Medium	High	High	Substantial	Substantial
Possible	Small	Medium	High	High	Substantial
Unlikely	Small	Medium	Medium	High	High
Practically impossible	Small	Small	Small	Medium	Medium

4 risk levels	
Small risk	safety is largely guaranteed
Medium risk	safety is partially guaranteed normal protective measures are required
High risk	safety is not ensured, protective measures are urgently required
Substantial risk	safety is not ensured, enhanced protective measures are urgently required

Figure 12.3: IATA risk matrix

Every day many reports are processed and the assignment of a risk level is a judgement based on experience. There is not time for a methodological approach to risk assessment for every event. This is not considered a shortcoming. Some airlines have a 3 by 3 matrix for risk assessment and as a reaction on the question, whether so few scales does not oversimplify too much, the answer was based on a practical approach: 'that you can't go wrong more than two scales', which means you are almost always nearly correct.

With an interval of several weeks, trend reports are made and the reports are assessed with a team consisting of safety officers and a representative from the aircraft type office, e.g., a flight technical pilot. The risk level might be reassessed with expert judgement and further possible actions on specific reports and possible trends are discussed.

The challenge in these meetings is to distinguish between noise and signal with respect to threats for safety. How can attention in a smart way be given to the events that have the potential to become incidents or even accidents? What are the signals that indicate a real threat and how can the evaluation of all the 100 to 200 reports per month in an average airline be managed?

In addition to risk, analysis reports are also categorised. The categories used to classify the air safety reports are a mix of genotypes and phenotypes. There is, e.g., a category with 'human factors' which is a genotype, inferred from data from the report, and a category with 'air proximities' (insufficient separation between aircraft) which is a phenotype, the consequence of something. The problems caused by mixing genotypes and phenotypes, such as mixing causes and effects, are discussed by Hollnagel (1993a).

Accident or Incident Investigations. The ICAO emphasises the importance of analysis of accidents and has published an annex, which is a strongly recommended standard, specifically aimed at aircraft accident and incident investigations. ICAO requests compliance with their annexes from their member states, which are 188 (in 2003) countries in total. The ICAO document with its annexes originates from 1946, and today ICAO annex 13 is used as the world standard for investigations conducted by airlines and the national aviation safety authorities such as the NTSB (National Transport Safety Board).

Annex 13 provides definitions, standards and recommended practices for all parties involved in an accident or incident. One specific advantage from following annex 13 is the common and therefore recognisable investigation report structure along with terms and definitions.

Annex 13 stresses that the intention of an investigation is not to put blame on people but to create a learning opportunity. To seek to blame a party or parties creates conflict between investigations aimed at flight safety, focussed on the *why* and *how* of the accident and investigations conducted by the public prosecutor aimed at finding *who* to blame and prosecute. Fear of prosecution reduces the willingness of actors in the mishap to cooperate and tell their stories. Conflicting investigation goals hamper the quality of the safety report and reduce the opportunity to learn.

For classifying an event as an accident the ICAO annex is used. The ICAO definition of an accident states that it is when a person is injured on board or when the aircraft sustains damage or structural failure that adversely affects the structural strength, performance or fight characteristics and would require major repair, except for contained engine failure or when the aircraft is missing. This definition does not lead to much confusion but the definition of a serious incident is more negotiable, see Figure 12.4.

Air Traffic incident
    A serious incident involving Air traffic Control
Technical incident
    Damage or serious cracks, corrosion etc found during maintenance or inspections
    Fires and false fire warnings
    Contained engine problems
    Failure of onboard systems requiring non-normal operating procedures
Ground incident
    Collision with cars, terrain etc.
Operational incident
    Not being a technical or air traffic incident
    Incapacitation of a crewmember
    Rejected take-off
    Exceeding operational limitation e.g. hard landing, over speed, low speed
    A significant error in fuelling or load sheet
    Critically low in-flight fuel quantity
    A significant deviation from track of flight level
    Landing at airport other than destination except for reasons of weather

---

Figure 12.4: ICAO definitions

Due to lack of support from the definitions, safety officers decide on the basis of their knowledge and experience how to classify an event. Based on ICAO event categorisation and or the risk level assigned, the flight safety manager and chief investigator will advise the director of flight safety whether to conduct a further investigation on this event.

Accidents are often investigated by the national transport safety authorities in collaboration with other parties such as the airline concerned, aircraft and engine manufacturers. An incident investigation in an airline is conducted by several investigators, of which often at least one is a pilot (all working for the airline) and often complemented by a pilot union member, also qualified as an investigator. Analysis of the collected accident data occurs often without a methodology but is based on experience and knowledge of the investigating team. This means that the qualifications of the investigators shape the resulting explanation of the accident. An investigation results in a report with safety recommendations aimed at preventing subsequent accidents by improving, changing or inserting new safety barriers. Responsibilities for the safety department stop when the report is delivered to line management. Line management has the responsibility and authority to change organisational or operational aspects to prevent re-occurrence of the accident by implementing the report recommendations.

It is repeatedly remarked that (with hindsight) it was clear the accident or incident could have been avoided if only the signals that were present before the accident were given enough attention and had led to preventive measures. Is this the fallacy of hindsight or are there better methods that can help with prevention?

Flight Data Monitoring and Cockpit Voice Recorder. Another source of data consists of the on-board recording systems in the aircraft that collect massive amount of data. The flight data recorders (the 'black boxes') record flight parameters such as speed, altitude, aircraft roll, flight control inputs etc. The aircraft maintenance recorders register system pressures, temperatures, valve positions, operating modes etc. The flight data and maintenance recordings are automatically examined after each flight to check if parameters were exceeded. This is done with algorithms that check, e.g., for an approach to land with idle power. This is done by combining parameters. Making these algorithms, which are aircraft specific, is a specialised job.

The cockpit voice recorder records the last 30 minutes of sounds in the cockpit. This is a continuous tape which overwrites itself so that only the last 30 minutes are available from the tape. The sound recording stops when a pilot pulls the circuit breaker to save the recordings or when the aircraft experiences severe impact during a crash. Data from the cockpit voice recorder are only used in incident or accident investigations and the use of information from this tape is strictly regulated by an agreement between the airline and the pilot union to preserve confidentiality of the pilots.

Quality and Safety. JAR-OPS (Joint Aviation Requirements for Operations) states that the quality system of an operator shall ensure and monitor compliance with and the adequacy of, procedures required to ensure safe operational practices and airworthy airplanes. The quality system must be complied with in order to ensure safe operations and airworthiness of the airline's fleet of aircraft. This statement is indicative of a perspective that safety and quality have a large overlap, and that safety is almost guaranteed as long as quality is maintained.

Quality management is currently regarded as a method of pro-active safety management. While accident and incident investigations are re-active, quality audits and inspections are pro-active. Quality audits compare organisational processes and procedures as described in the companies' manuals with the rules, regulations and practices as stated in, e.g., the JAR or IOSA. Company rules and procedures should comply with these standards and should be effective. The measure of effectiveness is based on a 'common sense' approach (remark by an IATA accredited auditor) and compared to best practices as performed by other airlines. Auditors are mostly not flight safety investigators and *vice versa*. Quality inspections are observation of people at work and the comparison of their activities with the rules and regulations as described by the company in the manuals. Audit observations and interpretations are thus clearly not black and white issues but are, to a certain extent, negotiable. Audits and inspections lead to recommendations for changes to comply with rules and best practices.

Airline Safety Data Sharing. Safety data is shared among airlines as reported in the GAIN (Global Aviation Information Network, 2003) work package on automated airline safety information sharing systems. Several systems for data exchange exist with the aim of learning from other airline's experiences. No single airline may have enough experience from its own operations for a clear pattern to emerge from its own incident reports, or may not yet have encountered any incidents of a type that other airlines are beginning to experience. In some cases, such as introducing a new aircraft type or serving a new destination, an airline will not yet have had the opportunity to obtain information from its own operations. Therefore, the importance of sharing information on both incidents and the lessons learned from each airline's analysis of its own safety data is also becoming more widely recognised.

When, for example, an airline has a unique event, the safety officer can search in the de-identified data for similar cases. If useful information is found, a request for contact with the other airline can be made to the organisation supervising the data. Via this organisation the other airline is requested if contact is possible. This data sharing has been very useful for exchanging knowledge and experience.

Safety Organisation Publication. Organisations like the FSF (Flight Safety Foundation), GAIN, ICAO organise seminars and produce publications on safety-related issues. Periodical publications are sent to member airlines where those papers are collected and may serve as a source of information when a relevant issue is encountered. Daily practice shows that due to work pressure little or no time is used to read the information given through these channels. The documents are stored and kept as a reference when specific information is required.

As a summary, an overview of the mentioned data sources is supplied in Figure 12.5.

Safety Management Targets

Even though airlines may not market themselves as 'safe' or 'safer', safety is becoming a competitiveness aspect between airlines. This can be seen from the several websites that compare airlines and their accident history. The disclaimers on these sites indicate that such ratings are just numbers and the numbers only give a partial picture of the safety level of an airline. The way data are complied could tell a lot about the validity of the ranking numbers but the general public might not be able to make such judgements. Still, there is a ranking and it could influence people's perception of safe and less safe airlines.

The term 'targets' is confusing if the number is higher than zero. Does an airline aim for, e.g., five serious incidents? No, in safety the word target has more the meaning of reference. This is what I understood when discussing this with safety managers, but still the word target is used.

Safety targets can be defined as the number of accidents, sometimes divided into major- and minor accidents, serious incidents and incidents and this number is indicative of safety performance. The definitions used for event categorisation are very important because this determines in which category an event is counted. As a reference for definitions, the ICAO annex 13 is used. These definitions are widely accepted but interpretations, as in which category an occurrence is placed, are often highly negotiable. A large list of event descriptions is published in annex 13 to facilitate these categorisation process discussions. The trap here is that events are compared with items on the list on phenotypical aspects, on how the event appears from the outside, while the underlying causes or genotypes may make the events qualitatively different. This makes the judgement for severity, what determines the classification, very negotiable.

Data source	Method of compilation	Interpretation and usage
Air Safety reports	Paper report sent by pilots	The rules require the submission of a report. Safety officer categorises and assigns risk level.
Investigations of accidents and incidents	Team of investigators need 3 to 9 months to complete the report and compile recommendations.	Recommendations are discussed with line-management, accepted recommendations must be incorporated.
Flight data monitoring	Automatic registration of flight and system parameters	Screened by computers, when outside bandwidth a specialist makes interpretation and can advise for further investigation.
Cockpit voice recorder	Automatic recording of last 30 to 90 minutes of sounds in the cockpit.	Only used in accident investigation under strict rules.
Quality audits	Regulations require audits done by qualified auditors	Recommendations must be complied with.
Airline data share	Automatic exchange of de-identified data	Database interrogated when specific information is required.
Safety organisation publications	Organisations send their members publications.	Publications browsed and stored.

Figure 12.5: Overview of data sources

The target value for accidents is zero, but some airlines distinguish between major and minor accidents and have a non-zero value for minor accidents. The target values for serious accidents and incidents are often based on historical data, and the number from previous year is then set as the target.

An airline's current safety status can be expressed in counts of occurrences. Occurrences are categorised in a fixed number of categories and are assigned a risk level and are published in management reports together with historical data to show possible trends. Quantitative indicators may, in this manner, lead to further investigations and analyses of categories with a rising trend while absent or declining trends seldom invoke further analysis.

Workload is high in a safety department as in other production related offices. Much work has to be done with a limited number of people and this results in setting priorities and shifting attention to noticeable signals such as rising trends, accidents and incidents. Weaker signals and investigations into factors higher in the organisation that could have negative effects may consequently get less attention than practitioners really want. For example, can we find threats due to a reduction of training, which have not yet been recognised in incidents? These are pro-active activities for which currently the time and knowledge is lacking. Daily administrative duties and following up 'clear' signals such, as air safety reports, absorb the available human resources.

On a regular basis, management information is complied from reports and data monitoring. The data presentation shows graphical counts of occurrences, risk classification and history. Text is added with remarks about high risk events, and there is other explanatory text written by the flight safety manager. A de-identified extract of such a report is shown as an example in Figure 12.6.

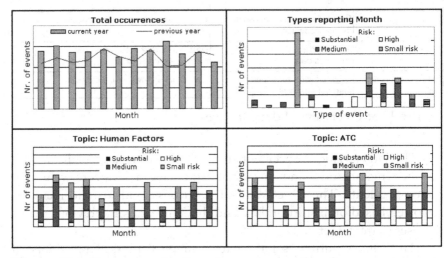

Figure 12.6: Example of safety data presentation for management

Upper management is very time constrained and safety information must be delivered in a simple representation. This (over)simplification is recognised, but 'that is the way the management world works'. These kinds of reports have to trigger upper management for requests for further investigation on the why and how on specific issues. Some reports for upper management compare actual counts compared to the safety targets. It would be interesting to know what meaning is assigned by upper managers to a value that is below the target. Is it only when a count exceeds the target that action is required? Much has been published on the problems of using counts as indicators (Dekker, 2003a).

Summary Safety Practices

This is a description of current airline safety management practices constructed from GAIN reports and communication with practitioners and responsible managers. Different sources of data show that the aviation community puts much effort into sharing information and trying to learn. Many data are of the same dimension, and manifestations of events and common underlying patterns are occasionally sought. Events are converted into numbers, classifications and risk levels and numbers are the main triggers for action from management. This is the way it has been done for a long time and practitioners and management lack resources to self-initiate substantial changes to safety management. They have to turn to science to supply practical diagnostic and sensitive approaches. When no alternatives are supplied, safety management will marginally improve and numbers will rule forever.

In the next sections the theoretical foundation of the current practices will be discussed and in the final section I propose a 'wish list' of what would be necessary to move forward in practice.

Models of Risk and Safety

Aviation is very safe, on average one accident in a million flights, so why should airlines try to improve this figure while they experience almost no accidents? Managers come and go about every three years, often too short to experience the feedback from their actions. Many years may pass, depending of airline size and flight frequency, before an accident occurs. From the manager's perspective it can be hard to justify safety investments; which have unclear direct results, especially in relation to cost-cutting decisions that have clear desired short term effects and are deemed necessary for economical survival. How do most accountable managers and practitioners perceive flight safety data, and what models of risk and safety actually shape their decision and actions?

From Data to Information and Meaning

As the description above shows there is much data available in a safety department. But is it the right data? Can safety be improved with this sort of data or is it just number chasing?

A critical factor for success is the translation from data to information and meaning. Collecting all the data points in a software database structure is not the same as interpreting data and trying to understand what the data points could mean for processes that may influence flight safety. Making sense of the data is complicated by classification methods and current practices are not able to deal with this.

First, all events are categorised and problems arise when categories are mixed in the sense of antecedents and consequents of failure as an event description. As said, Hollnagel (1993a) explains how classification systems confuse phenotypes (manifestations) with genotypes ('causes'). The phenotype of an incident is what happens, what people actually did and what is observable. Conversely, the genotype of an incident is the characteristic collection of factors that produce the surface (phenotype) appearance of the event. Genotypes refer to patterns of contributing factors that are not observable directly. The significance of a genotype is that it identifies deeper characteristics that many superficially different phenotypes have in common.

The second issue is that of classification of occurrences. An often complex event in a highly dynamic environment is reduced to one event report. Shallow information, no contextual data and lack of method to assign a category make categorisation a process that is greatly influenced by the personal interpretation of the practitioner doing the administration. This also goes for the risk level that is assigned to an event and used to rank the events.

Third, the requirement from most of the database software is to put events in a single category. Most software tools cannot handle events in different categories at the same time. Counting gets difficult when some events are assigned to more than one category and counting events per category is regarded as valuable information.

Management is presented with numbers and some qualitative descriptions of high risk events. An initial quantitative approach on occurrences is often used and a further qualitative approach is postponed to the point where a trend, a rise in occurrences in a specific category, is observed. When a trend is observed an analysis of this data is started and this is then the initiation of a more qualitative analysis where data is interpreted and assigned a meaning.

All the mentioned re-active practices are assumed to be useful based on the 'iceberg' theory, which basically means that preventing incidents will prevent accidents. But is this still a valid assumption?

Learning from Accidents

Accident and incident investigations are executed to create a learning opportunity for the organisation. As a result, from the investigation a report is published with safety recommendations that, according to the investigation team, should be implemented to prevent re-occurrence of the mishap. The responsibility of the safety department ends when the report is published; thereafter it is up to the line management (e.g., fleet chiefs) to consider the recommendations in the report. Line management has the authority to interpret the recommendations and can reject or fully implement the changes suggested by the investigators. First, often recommendations can have a negative effect on production, and thus on cost. Implementation may include extra training for pilots or changing a procedure or limitations of equipment. Line management will make trade-offs between these conflicting goals and it is not often that a recommendation is rejected. Second, failure is often seen as a unique event, an anomaly without wider meaning for the airline in question. Post-accident commentary typically emphasises how the circumstances of the accident were unusual and do not have parallels for other people, other groups, other parts of the organisation, other aircraft types.

One airline is experimenting by involving line management when the recommendations are defined. Having early commitment from line managers, to comply with the recommendations that they themselves have helped to define, can increase acceptance of the final recommendations when the report is published. Without acceptance of the suggested changes the time and money invested in executing the investigation is wasted, since no change, or learning has occurred. A downside of involving line managers in this early stage might be that only easy-to-implement changes are defined as recommendations and that the harder and more expensive changes are not accepted. It was suggested to use the acceptance of recommendations as a quality indicator for the report. Since costs of implementation play an important role in managers' acceptance of recommendations it might be more suitable to use acceptance of the suggested changes as a measure of quality of the learning organisation.

The quality management system in an organisation checks if the accepted recommendations are indeed implemented. There is, however, no defined process that analyses if the recommendations had the intended effect. Only if new events are reported that seem to relate to the installed changes can a connection be made and an analysis done of the 'old' recommendations.

Much effort is put into investigations and after a few weeks and months operations go back to 'normal'. The investigation is still in progress and only after many months or sometimes after two or more years, is the final report published. This delay can be caused by factors such as the legal accountability of the parties involved and their subsequent withdrawal of the investigation, or a lack of resources in the (national) investigation bureaus. The eagerness to use the final recommendations a long time after the accident is considerably reduced. The possible learning opportunity was not effectively used and the system's safety is not improved.

Quality versus Safety

Current practices in safety management build on the view that quality management is the pro-active approach to preventing accidents. Quality management focuses on compliance with procedures, rules and regulations of people and organisations in their activities. Audits and inspections are done to confirm this compliance and, in a high quality organisation, few deviations from the standards are observed. This raises the question of whether high quality organisations are safer than lower quality organisations. The approach that quality guarantees safety builds on a world view that compliance is always possible, that all procedures are perfect, that no failures outside the design occur, that all performance is constant, and that no constraints exist which limit the ability of people to have full knowledge certainty and time to do their work. This view explains that accidents are caused by deviations from the approved procedures, rules and regulations, thus quality and safety have a complete overlap.

A more realistic view is that the complexity and dynamics of our world are not so predictable that all rules, regulations and procedures are always valid and perfect. Due to coincidences, people have to improvise, because sometimes conditions do not meet the specifications as stated in the procedure, and trade-offs have to be made between competing goals and this is normal work. In this view there is far less overlap between quality and safety, it is assumed that procedures, rules and regulations are as good as they can be but at the same time their limitations are recognised.

Quality as a pro-active approach to safety is a limited but research and science has not yet delivered practical alternatives.

What Next? From Safety to Resilience

Having given an interpretation of current practices and having pointed at some problematic issues, here are some questions to the researchers and scientists:

> Should data collection only be done from the 'sharp end', the operators or should this be complemented by data gathering from other cross-company sources such as maintenance, ground services and managerial failures?

> How can the above-mentioned data sources be combined to provide a wider perspective on safety and risk?

> Current practices use only flight operation related data sources. Maintenance event reports are not evaluated outside the maintenance department. Managerial accidents, such as failed re-organisations are not investigated like operational events but all these different types of failures occur in the same organisational context so common factors can be assumed.

> What sort of data should be collected from operational events with, e.g., an air safety report?

Current practices show reports with descriptions of events and only a few contextual factors. Classification and risk assessment are done by a flight safety officer.

> How can it, with the evaluation of events, be determined whether the event is signal or noise in relation to safety or risk and whether putting resources in the investigation is justified?

Operational events (not yet determined if it is an incident) are reported to the safety department. Safety officers and management have to decide in a short time frame (the flight is, e.g., postponed until the decision is made) whether an investigation is deemed necessary.

Safety reports are evaluated periodically and, based on their assigned risk level, further investigation or other follow up actions may be decided.

> What kind of data should be presented to management to keep them involved, willing to follow-up on findings and supportive for actions such as investigations and risk analysis?

Experienced and knowledgeable safety officers have a feeling certain indications, e.g., an increased numbers of technical complaints about the aircraft, point to increased risk. But in the 'managers' world' it requires 'hard facts' or evidence to make (upper) management react to these signals. Currently, owing to the immense work pressure on management, data are presented in simplified form that takes little time to evaluate.

> How can risks be predicted and managed if, e.g., training of pilots is reduced, procedures changed and new technology introduced?

Judgements about possible risks are made 'during lunch' since no methods are available to predict the consequences of changing procedures and if, e.g., additional training is required when the procedure to abort a take-off is changed. What kind of problems can be expected and how should these be handled when, e.g., new technology such as a data link, is introduced. Nowadays, engineering pilots and managers with much operational experience but limited human factors knowledge try to manage these issues.

How can a common conceptual and practical approach to reporting, investigation and risk prediction be developed?

The three issues of reporting, retrospective investigations and pro-active risk prediction have no commonalities whatsoever. These isolated approaches do not reinforce each other and valuable insight may be missed.

Chapter 13

Taking Things in One's Stride: Cognitive Features of Two Resilient Performances

Richard I. Cook
Christopher Nemeth

> *"If you can keep your head when all about you*
> *Are losing theirs and blaming it on you..."*
> Rudyard Kipling, 1909

Introduction

Resilience is a *feature* of some systems that allows them to respond to sudden, unanticipated demands for performance and then to return to their normal operating condition quickly and with a minimum decrement in their performance.

Our approach to resilience comes from the perspective of practitioner researchers, that is, as people engaged directly with sharp-end work and sharp end workers. Regarding human performance in complex systems, we take two things for granted:

- The many-to-many mapping that characterizes a system's goal – means hierarchy relates the *possible* to the *necessary* (Rasmussen, Pejtersen & Goodstein, 1994).
- Human operator performance in these systems is encouraged, constrained, guided, frustrated, and enacted within this hierarchy (Rasmussen, 1986).

The location of the system's operating point is the result of a balance of forces (Cook & Rasmussen, 2005) and tracing the movement of that point over time maps the history of that system's safety.

This chapter describes some of the cognitive features of resilience. We seek to prepare the ground for discussions of resilience and its characteristics using two examples. While the following accounts of two events are both drawn from the domain of healthcare, they have few overlapping features. One occurs over a very short period of time, is a highly technical problem, and has a few actors. The other spans a half a day, encompasses a large socio-technical system, and involves hundreds of people. Despite their differences, the two events demonstrate resilience. To be more precise, each is a *resilient performance*.

Example 1: Handling a 'Soft' Emergency

Setting. The Department of Anesthesia (DA) in a major urban teaching hospital is responsible for management of the Surgical and Critical Care unit (SACC). The unit includes six activities in the hospital complex that require anesthesia services. Cases are scheduled for the outpatient clinic (SurgiCenter), the inpatient operating rooms (IOR) or four other services, such as radiology. The most acute cases are treated in the IOR. Cases vary from cardiology to neurology to orthopedics, transplants and general surgery. In each case, staff, equipment, facilities, and supplies must be matched to meet patient needs and surgeon requirements.

The DA employs a master schedule to funnel cases into a manageable set of arrangements. Completed each day by an anesthesia coordinator (AC), the schedule lists each case along with all needed information. The finished master schedule, which is distributed throughout the organization at 15:30 each day, is the department's best match between demand and available resources. One hard copy is posted at the coordinator station that is located at the entryway of a 16-operating room SACC unit. There, the AC and Nurse Coordinator (NC) monitor the status and progress of activity on the unit as the day progresses. Surgeons reserve OR rooms by phone, fax, e-mail, and in-person at the coordinator station. This flow of bookings is continuous and varies according to patient needs and surgeon demands. Because the flow of demand for care is on-going, surgeons continue to add cases through the day. These are hand written onto the master schedule 'add-on list.'

Cases are either scheduled procedures or emergencies. Each surgeon estimates case severity as either routine, urgent or emergency. If a room is not available, a case is pre-empted ('bumped') to make resources available. While the bumping arrangement works, the staffs prefer not to use it if possible. That is because bumping causes significant changes to the way staff, rooms and patients are assigned, which the unit would prefer to avoid. Officially, the potential loss of a patient's life or vital organ would be an emergency; and its declaration requires that resources be freed up and made available for that case. In reality, there are very fine gradations of what constitutes an emergency. Because of that ambiguity, it is possible for a surgeon to declare a case 'emergent' in order to free up resources for his or her case. In the instance of cases that are declared emergent, it is up to the AC to evaluate the case. Anesthesia and nursing staff members use the term 'soft emergency' to those cases that are declared yet have no apparent high risk. It is up to the AC to 'read' each declaration to ensure that there is an equitable match between demand and resource.

It is Friday at mid-day and surgeons are attempting to clear out cases before the weekend. The master schedule add-on list is 2.5 pages long. One page is more routine. The chief anesthesiology resident speculates that there may be a backlog because admitting was suspended on Thursday due to no available beds (the queue 'went solid'). At around 11:30 the Anesthesiology Coordinator fields a phone call to the AC. The following sequence of events occurs over about ten minutes from 11:30 to 11:40. Portions of the dialog (Nemeth, 2003) that occur at the Coordinator station provide a sense of what is occurring.

Observations. Operating room number 4 is the day's designated bump room. The sequence begins as the AC explains to Surgeon 2, whose next case is scheduled in the bump room, that an emergency declared by Surgeon 1 is bumping him and, therefore, that his next case may not be done today.

AC: [on phone with Surgeon 2] "I don't think we can do it today. We can book it and see if we can fit it in." [pause]

"You're in the bump room, number 1 [in line to be bumped]."
[pause]

"He [Surgeon 1] wants to do it." [pause]

"I don't make those decisions." [pause]

"You'll have to talk to him [Surgeon 1]."

Nurse1 [from Room number 4]: "Are we being bumped?"

AC: "Yes, you're being bumped."

Nurse: "How do they decide which rooms get bumped?"

AC: "The bump list."

Nurse: "Doesn't it make sense to bump by the severity of the case?"

AC: "It depends on the surgeons to make those decisions."

AC: [to NC] "Clean it up and let's set it up. Ask the person bumping to talk to the surgeon who is being bumped."

--break--

NC: "[O.R. number] Six is coming out."

AC: [to NC] "We can put [Surgeon 1] in there."

   [Calls on the phone and waits, no answer.]

   [to NC] "We can put [Surgeon 2] in [OR number] 4 and [Surgeon 1] in [OR number] 6."

   "Let me page [Surgeon 1]."

--break--

   [to NC] "You're planning to put something in [OR number] 6."

   [points to a case on the add-on list]

   [pages Surgeon 1]

   [calls Surgeon 2]

--break--

Nurse1:[while walking past station] "Are we being bumped? Is this for sure now?"

AC: [to Nurse1] "Do the scheduled case."

--break--

AC: [to Surgeon 1] "*Surgeon 1?* Room 6 is coming out." [pause]

--break—

AC: [to Surgeon 2] "*Surgeon 2?* I just talked to *Surgeon 1* and he doesn't mind waiting 45 minutes [for OR number 6 to be setup and the emergency case to start]."

The surgeon had called to declare an emergency in order to have his skin graft case (an add-on procedure) assigned to an operating room when none were open. Opening a room would require a scheduled case to be bumped. The AC negotiated an arrangement in which the surgeon accepted a 45-minute delay until a room became available. The arrangement avoided bumping a procedure. The AC reflected on the process shortly afterwards:

> When they declare emergency I don't have anything to do with it. We let them work it out. I struck a deal by getting [Surgeon 1] to wait 45 minutes and was able to put him into a room that was coming open. That avoided having to bump a procedure.

This particular AC has a reputation for many capabilities, notably for his expertise in coping effectively with demands for resources. In this case, he confronts a situation that formally requires a particular course of action – the use of the bump room. Although he is prepared to follow this disruptive course, he does more than the formal system requires. He manages the problem expectantly, discovers an opportunity, and presents it to the parties for consideration, all the while reminding all the parties of the formal requirements of the situation. He manages to avoid confrontation by avoiding a controlling role. Instead, he invites surgeons to work out the resource conflict between them. By offering the bumping surgeon a viable alternative and inviting him to accept it, he relieves the tension between resource and demand. He averts a confrontation, making good use of available resources. He understands the surgeons' personalities, the nature of the kinds of cases at issue, the pace of work in the operating room, and the types of approaches that will be viable. He evaluated the skin graft as a 'soft emergency', that is, a case that could be delayed but not put off indefinitely.

The evolution of responses that the AC gave through the brief period indicates how elastic the boundaries are between what exists and what might exist. Circumstances changed from a demand for unavailable resources, to the potential for an arrangement, to the reality of an arrangement. As he saw that evolving, he adjusted his responses in order to set actions in motion so that resources were ready for use when needed (e.g., "Clean it up and let's set it up," and "Do the scheduled case").

The episode is subtle, small, and embedded in the ordinary flow of work in ways that make it almost invisible to outsiders. The AC employed complex social interactions and various kinds of knowledge to reorder priorities. The event's significance lies in the absence of disruption that the AC's efforts made possible in the face of a sudden demand for resources.

Example 2: Response to a Bus Bombing

Setting. At 07:10 on Thursday morning, 21 November, a suicide bombing on #20 Egged bus occurred while the bus was in the Kiryat Menahem neighborhood of Jerusalem. The bomb was detonated by a 22 year old Palestinian from Bethlehem who had just boarded the bus and was standing near the middle of the bus at the time of detonation. The explosive device included a variety of shrapnel objects; typically these are ball-bearings, small bolts or screws, or nails. Early reports from the scene said that seven people had been killed. Two persons evacuated from the scene died in hospital on the first day. Forty-eight other people were wounded; twelve people had died of their wounds by the end of the day (eleven plus the bomber).

Bus attacks by suicide bombers have fairly monotonous features. They occur during the morning rush hour because ridership is high at that time. Bombers board busses near the end of their routes in order to maximize the number of people in the bus at the time of detonation. They preferentially board at the middle doors in order to be centered in the midst of the passengers. They detonate shortly after boarding the bus because of concern that they will be discovered, restrained, and prevented from detonating. They stand as they detonate in order to provide a direct, injurious path for shrapnel. Head and chest injuries are common among seated passengers. The injured are usually those some distance away from the bomber; those nearby are killed outright, those at the ends of the bus may escape with minor injuries. The primary mechanism of injury of those not killed outright by the blast is impaling by shrapnel. Shrapnel is sometimes soaked in poison, e.g. organophosphate crop insecticides, to increase lethality.

Traveling to the Hospital. A senior Israeli physician (DY) heard the detonation of the bomb at 07:10 and drove with one of the authors (RC) to the Hadassah Ein Karem hospital. It is routine for physicians and nurses to go to their respective hospitals whenever it is clear that an attack has taken place. Within 15 minutes of the detonation, FM radio broadcasts included the bus route name and location at which the explosion had occurred and provided early estimates of casualties. DY did not use his cell phone to call the hospital for information, explaining that there was no reason to do so, that he could not participate effectively by telephone, and that the senior staff person at the hospital would already be managing the response. DY explained that this was a 'small' attack and that the fact that the operating room day had not commenced at the time of the detonation meant that there were ample staff and facilities to handle whatever casualties would appear.

Ambulance drivers are trained to go to the site of a mass casualty without direction from the dispatch center. The dispatch center triages ambulance hospital assignments once they are on scene and loaded with patients. Emergency personnel describe this approach as 'grab and go.' This leads to rapid delivery of patients to hospitals where definitive care may be obtained. It decompresses the scene so that police forensic and recovery operations can take place unhindered by the presence of injured people. It also reduces the risk of further injuries from secondary attacks.

At the Hospital. As the largest modern hospital in the Mid-East, Haddassah Ein Karem could be expected to be busy and crowded in the morning of a regular workday but it was especially so at this time. Although the detonation had been less than an hour before, people were already arriving to search for family members. The hospital had already dedicated a large meeting room for the families and friends, and hand-lettered signs were posted indicating the location of this room. In the room were trained social workers and physicians. These personnel had donned colored plastic vests with English and Hebrew print indicating that they were social workers. The room itself was supplied with tables and chairs, a computer terminal connected to the internet, and water and juice. This room had been selected, provisioned, and manned within 30 minutes of the detonation and in advance of the first families arriving at the hospital.

Staffing for the social work activities is flexible and it is possible to recruit additional support personnel from the hospital clerical and administrative staff. The goal of the social work staffing is to provide sufficient numbers of staff that there is one staff person for each individual or family group. Once attached to the individual or family, the social worker focuses narrowly on him or them for the duration of the emergency.

To the Operating Rooms. DY went immediately to the operating room desk for a brief conversation with the four or five anesthesiologists and nurses present there. He then manned the telephone at that location. He checked an internet terminal on the desk for the latest information from the scene and took a telephone call regarding a casualty being transferred in from another hospital. The apparent death toll was seven or eight and there were thought to be forty to fifty wounded. The wounded had already been dispersed to hospitals in the city. This triage of cases is managed by the ambulance drivers and dispatchers in order to distribute the casualties efficiently. In particular, the triage is intended to move the most seriously injured to the main trauma centers while diverting the minor casualties to smaller hospitals. Coordination to free up intensive care unit beds for the incoming casualties had begun.

Two severe casualties had been taken directly to the operating room, bypassing the trauma room. Operating rooms are used as trauma rooms for expediency: cases that will obviously end in the operating room bypass the trauma room and are taken by a team directly to the OR. This speeds the definitive care of these patients and preserves the three trauma room spaces for other casualties; observed activities were considerably faster and involved fewer people than in many US trauma centers.

To the Emergency Room. The emergency room can be reached by driving into a square that is approximately 80 meters across and surrounded on three sides by hospital buildings. Ambulances enter along one side of the square, discharge their patients, and proceed around the square and exit the area. Security personnel had erected a set of bicycle rack barricades to create a lane for the ambulances. This was necessary because the television and newspaper reporters and camera crews were present in force and crowded into the area that is normally reserved for the ambulances.

A hospital spokeswoman was present and patiently repeated the minimal information that was available. Members of the press were intrusive and persistent and were the most visibly keyed-up people observed during the day. The hospital spokeswoman was remarkably patient and quite matter-of-fact in her delivery. The hospital's director of media relations was standing off to the side, uninvolved in continuous press briefing but monitoring the activities.

Arriving families were directed to the emergency room where they were met by social workers. Triage of casualties to multiple hospitals and the transfer of the dead to the morgue happens before identification. In many instances, casualties and the dead are not readily identifiable by medical personnel. This is particularly true for those who are dead or severely injured because of the effects of blast and fire.

Following a bombing, people begin by attempting contact bus riders by cell phone. If there is no answer or if the rider does not have a cell phone, family members go to the nearest hospital. Social workers at that hospital will gather information about the person being sought. All such information is shared via the internet with the all the other hospitals. Similarly, hospitals collect identifying information about the patients they have received, including gender, approximate age, habitus, and clothing. These two sets of information are then collated by social workers at each hospital. This helps the medical staff identify who is being treated and helps the families determine the location of the person they are seeking. Pooling the information makes it possible quickly to refine a casualty list and identify individuals. Police use this information from the hospital at the morgue formally to identify those who have died. In this case, a complete casualty list was available within about 6 hours of the bombing. The next morning's newspapers provided names and pictures of all the dead including the suicide bomber.

The Israelis place a high premium on providing family members with access to casualties. The author observed several sets of parents and siblings being brought into the preoperative holding area or the emergency room to be with the casualty. In each case, the family members were accompanied by at least one social worker.

For each person who is injured, as many as four people may be expected to come to the hospital. Parents, grandparents, siblings, and children may arrive simultaneously or in waves. The management of this crowd of people, who are all under intense psychological stress, is a major undertaking that requires substantial resources.

Care of a Severe Head Injury. An ambulance delivered a young female patient with a serious head injury to the Haddassah Ein Karem emergency receiving area. This patient had initially been taken to another hospital and was now transferred to this center where neurosurgeons were located. While the ambulance attendant provided a detailed description of the nature of the suspected injuries, medical staff took the patient to the trauma room performed resuscitation, and took physiological vital signs. The patient was found to have multiple, minor shrapnel wounds but no apparent major trauma other than a penetrating head injury. She had already been intubated and was being mechanically ventilated. She was in a cervical collar and secured on a transport backboard. All the elements of the Advanced Trauma Life Support (ATLS) protocol were observed: a complete examination was made, the cervical spine and chest were x-rayed and the C-spine cleared, and blood samples were sent for evaluation. Verbal exchanges in the trauma room were exclusively technical and the tenor of the conversation was relaxed. A senior radiology physician performed an abdominal ultrasound examination. The blood gases were returned within five minutes and slight adjustments to mechanical ventilation were made. The patient's vital signs were stable and the patient was mildly hypertensive. A negative pregnancy test result was returned at almost the same time. Narcotics and neuromuscular blocking drugs were administered.

Much of the paperwork that is completed during routine operations is deferred or abandoned entirely during a mass casualty response. A notable exception is the blood type-and-cross match paperwork, which is conducted using the routine forms and procedures. DY explained that abandoning some documentation was necessary to allow the critical documentation to be completed. Successful blood transfusion depends on accurate documentation.

A computerized tomography (CT) scan was planned and the patient was taken on a gurney to the CT scanner, while connected to a pulse oximeter, electrocardiograph (ECG), and non-invasive blood pressure monitor. The senior anesthesiologist, an anesthesiology resident, and a neurosurgeon accompanied her as a staff security person cleared a path through the hospital corridors. When the team and patient arrived in the CT room, they used a small, compressed gas powered ventilator that is permanently installed in the CT scan room, to continue mechanical ventilation. As the CT scan was made, the chief neurosurgeon arrived and, together with a radiologist, read images at the console as they were processed.

The decision was made to perform a craniotomy, which is surgery that involves opening the skull, and the team transported the patient to the operating room. The neurosurgeon received and answered a cellular telephone call about another patient as he pulled the gurney. While waiting for an elevator, the patient's parents came to the gurney, accompanied by a social worker. One of the transporting physicians spoke with them about the known injuries and the expected surgical procedure. The parents touched the patient, and looked under the blankets at her legs and feet.

The attending anesthesiologist, a scrub nurse, and a circulating nurse were waiting in the OR for the surgical team and patient to arrive. The anesthesiology resident provided an update on the patient's condition. After transfer to the operating room table, the patient was positioned, a radial arterial line was started and the head prepared for a craniotomy. The craniotomy began one hour after the detonation.

Return to Normal. Although several operating rooms were occupied with procedures that were related to the bombing, the remainder of the operating rooms had begun cases from the regular daily list. Efforts to resuscitate a trauma victim continued in one room. Several other patients with less serious injuries were in the preoperative holding area. No further direct casualties were received.

The author met briefly with anesthesiologists, surgeons, and nurses in the operating room corridor and then visited the emergency room and the family room. Many of the families had dispersed to other areas in the hospital as the patients received treatment, were admitted to wards, or were being discharged.

DY noted that the team generally holds a brief informal discussion of lessons learned immediately after a mass casualty response. The value of these discussions, he said, has fallen off substantially as the handling of such events has become routine.

Analysis

We propose the cases as examples of resilient *performances* and these resilient performances as *evidence* of the presence of a resilient system. Systems may be resilient but we do not know how to detect resilience short of the direct observation. This is because resilient performances occur in the face of sudden, unanticipated demands. *At present, the only strong evidence of resilience that we can identify is the presence of these resilient performances.* They are made *possible* by the configuration of the workplace, incentives, and social structures that surround them. This does not lessen our regard for the critical role that is played by the participants' attention, knowledge, assessments, and decisions. Confronted with a sudden demand, we observed the practitioners respond purposefully and appropriately. It was clear from their actions and speech that they recognized the demand and its significance. Their reaction responded directly to the demand. Their reaction was also temperate and tempered by other considerations. These resilient performances depend heavily, but not exclusively, on particular individuals and on the particular expert cognitive abilities that they bring into play. Through their experience, intentions, and judgement as events evolve, practitioners invest cognitive effort to maintain successful performance in the face of unexpected challenges. *Cognition creates what we observe as resilient performance.*

Unusual but Not Unknown: The Roles of Deep Domain Knowledge

Events that provoke resilient performances are unusual but not entirely novel. The practitioner in the soft emergency case knows that an 'emergency' can have different degrees of criticality. The experience of being called on to respond to a sudden demand is common enough that the practitioner has developed expertise in coping with such demands. Similarly, in the bombing case, the practitioners know from the earliest description of the event what sorts of wounds to anticipate, how patients will arrive, and where the bottlenecks in response are likely to be. They also know what resources they can call on and the established processes of dealing with this sort of event. In both examples, the practitioners use their knowledge of how operational tempo plays out, and what pathways can be exploited to respond.

Return to Normal Operations

Resilient performances include a response to a sudden demand *and* the return to 'normal' operations following that response. Concern for routine operations shapes the planning and execution of the response to the sudden demand. In the 'soft emergency' case, the response is fashioned in order to have a minimal effect on routine operations. This is a very practical concern. Routine system functions are important and expensive resources such as the OR are usually operated near, or at, full workload.

In the 'soft emergency', the anesthesia coordinator is able to foresee an opportunity to meet the new demand within the stream of activity. He knows that emergencies vary in criticality. Some require immediate response but others can tolerate a delay. He queries the surgeon to see whether this case can be delayed. He also examines current operations in order to see whether an operating room will, or could, become available.

In the bus bombing response, the response to sudden demand is continually assessed and normal work resumed even as the casualties from the bombing are being cared for. The resilient system is an inherently valuable resource so that interrupting its normal function generates costs and it is important to resume that normal function as smoothly and quickly as possible.

The concern for routine operations in the midst of sudden demand distinguishes resilience from other types of reactions. In both of these cases the system responds to a need without entirely abandoning routine operations. Normal operations continue in the face of the sudden demand. The response to the demand includes efforts both to gauge its effect on normal operations and to devise the response so as to disrupt those operations as little as possible.

Defining Successful Performance

Performances might be resilient without being successful, at least in the conventional sense. There are ways that these cases could have turned out better. If there were no delay in treating the burn patient, if the head injured patient had been taken directly to the main hospital instead of an outlying one, it is possible that these patients would have done better. Even so, the activities of these practitioners qualify as resilient performances. What, then, is the relationship between resilience and success? Can performances be resilient and unsuccessful? Are successful outcomes evidence of resilience?

Some sudden demands, such as large scale natural disasters, might be so overwhelming that any plan of action is destined to fail. In such situations, just returning to normal in the wake of an event could count as a resilient performance. Indeed, some of the more compelling examples of resilience are those in which the people and processes are restored after terrible outcomes. This is often evident from the small signs of normal activities that resume in the wake of a catastrophe. Abandoning attempts to meet the sudden demand and instead reserving resources to reconstitute the system after the demand has passed could, in our view, qualify as a resilient performance. But in each instance we note that resilient performance is not present because of a certain form nor is it absent if the outcome is undesirable. *Instead, the resilient performance itself involves either a tacit or explicit redefinition of what it means to be successful.*

Changing the Definition of Success

These examples and other cases of resilient performance involve a *shift in goals and redefinition of what constitutes success.* To perform resiliently, practitioners necessarily defer or abandon some goals in order to achieve others. Negotiating trade-offs across multiple goals is common in complex systems operations and it should not be surprising to find this type of activity in resilient performance. In these cases, however, the trade-offs occur at relatively high levels within the goal – means hierarchy. Rather than minor trade-offs, the sudden demand presents opportunities for a significant failure. The sudden demand creates a conflict that requires practitioners to refer to high-level goals. The resilience in resilient performance is the *degree to which the operators are able to handle the disruption to the usual goal – means features of their work.*

Traversing the Goal – Means Hierarchy

Rasmussen's goal – means hierarchy represents the semantic structure of the domain of practice. During operations, practitioner cognitive performances map a set of transitions through this hierarchy. In routine operational situations, the transitions are likely to map a small space at low levels of abstraction as workers perform 'routine' activities and achieve 'routine' results. Cognition at these relatively concrete levels of the hierarchy does not mean that the higher-level goals are not active. It is only that worker cognition does not engage high-level goals because the work activities do not produce conflicts.

Reference upwards in the goal – means hierarchy occurs when conflicts arise between goals that are located at the same level. The need to refer upwards in the hierarchy is brought about by the presence of conflicting goals. Most operational settings are set-up so that conflicts are rare. (If they were not rare, their operations would be very inefficient!)

Resilient performances are remarkable because the sudden demands that provoke them generate the need for practitioners to engage quite high levels of the goal – means hierarchy. A resilient performance is a cognitive performance that ably traverses the hierarchy in order to resolve a conflict. Within Rasmussen's goal – means hierarchy, a sudden demand will register as the appearance of a challenge to a goal high up in the hierarchy. The appearance of this challenge leads to the rapid working out of various trade-offs at lower levels as the means to cope with this challenge are sought by workers. This working out appears to observers as a resilient performance.

Taking it in One's Stride

Our colleagues regard resilience as a desirable and distinct feature of some systems and we agree with them. The skill and expertise that people use to cope with the challenges they face is frequently impressive and sometimes astounding. The robustness of resilient performances is often remarkable. The response to a bus bombing is a notable example because it is so vital and so dramatic. But the handling of a 'soft emergency' by the anesthesia coordinator is also remarkable, in part because it is so graceful and subtle.

When they are confronted with potentially disruptive demands, the people involved continue working. They direct their attention to the emerging threat to a high-level goal. They sacrifice lower level goals to meet the challenge but do so in ways that preserve the system. They monitor the escalating involvement with the threat and track the ebb and flow of operational responses. They strike a balance between committing resources to handle immediate needs and reserving resources for future ones. They deliberately accept smaller losses and failures in order to preserve opportunities for more valuable gains and successes. They anticipate the end of the threat and organize the resumption of 'normal' activities.

Those who are familiar with the work on cognition in process control settings will recognize that the litany of practitioner actions in a resilient performance is very similar to those of ordinary operations. This leads us to conclude that resilience is not a discrete attribute but rather an expression of some systems' dynamic functional traits.

Resilience can be large or small scale, depending on how high up in the goal – means hierarchy the sudden demand's challenge appears. The monotonous, familiar demands of a static world impose no need for resilience. The goal – means hierarchy for that domain may be complex. Practitioner cognitive tasks, though, are firmly bound to only the lowest levels of the hierarchy. Resilience is what we need to cope with the sudden appearance of a demand that puts a high level goal at jeopardy. Observers recognize resilient performances when the threat is abrupt and the goal put at risk is high in the hierarchy. The deeply resonant resilience that is so beautifully portrayed in the system that responded to the bus bombing is a quantitatively, but not *qualitatively*, different performance from that seen in the 'soft' emergency case. In both instances practitioners took the unusual event *in their stride* without missing a step.

Conclusion

We propose that resilient performance is empirical evidence of resilience. The need for resilient performance is generated by the appearance of a sudden demand for resources – a demand that threatens a high level goal. What distinguishes resilient performance is the fact that practitioners are able to move through the goal – means hierarchy to address the threat. In this formulation, *cognition* is the critical factor in resilient performances and the central feature of what it means for a system to possess the dynamic functional characteristic of resilience.

If our conclusions are correct, then research on resilience will likely be some combination of three themes. The first is research on cognition – including distributed cognition – in demanding situations.

The second is research on the explanation of goal – means hierarchies in naturalistic settings. The third is research on the characteristics of sudden demands for resources and the reactions that they evoke. Understanding resilience is likely to depend almost entirely on assembling research approaches that explore the interactions among these three themes.

There is good reason to be encouraged about the prospects for successful research on resilience. It is certainly difficult to study dramatic events such as the bus bombing we described. But resilience is also found in the everyday operations of complex systems. It is part of the mundane, messy details of real work in the hospital and, we are certain, elsewhere. Although the scope of the threat in these two cases is very different, the mechanisms of resilience are, we believe, the same. The results that we obtained from the study of the work of anesthesia coordinators offer a means to understand the dramatic response to large scale disasters such as bus bombings.

Chapter 14

Erosion of Managerial Resilience: From Vasa to NASA

Rhona Flin

Organisational resilience is defined in this volume as 'the characteristic of managing the organisation's activities to anticipate and circumvent threats to its existence and primary goals. This is shown in particular in an ability to manage severe pressures and conflicts between safety and the primary production or performance goals of the organisation' (Hale & Heijer, Chapter 3). While the organisational or systems level of analysis offers one perspective on resilience, for practical application, it is also necessary to understand how the concept can be decoded to a behavioural level of analysis. What do workers or managers actually do differently in more resilient organisations?

If the above definition is translated to a managerial level, then the essence of resilience appears to be the manager's ability to deal with conflicts between safety and performance goals. Woods conceptualises this as the skill in making 'sacrificial decisions' (Chapter 2). In order to reduce the risk of accidents, managers have to recognise threats to safety and rebuff or circumvent these while minimising threats to productivity. In cases where this component of resilience fails – managers do not detect the threats or they recognise the threats but fail to take the requisite action.

Post-hoc analyses of accidents have well documented limitations but they do offer salient illustrative material. A number of authors in this volume have referred to the space shuttle explosions, but much earlier accounts of transportation disasters also reveal erosion of managerial resilience.

Vasa to Columbia

In 1625, the Swedish King, Gustavus Adolphus, was building an empire around the Baltic Sea, and to this end, ordered several new warships. Among them was the Vasa, which was to be built in Stockholm by a Dutch shipbuilder. His experience was essential as the Vasa was to be the mightiest warship in the world, armed with 64 guns on two gundecks. The finished battleship was a magnificient vessel with three tall masts and her wooden hull decorated with elaborate carvings of lions' heads, angels and mermaids. On Sunday August 10th 1628, she set sail from Stockholm on her maiden voyage, watched by a large crowd of spectators, including foreign diplomats (Figure 14.1). After a few minutes, she heeled over and sank to the bottom of Stockholm harbour. Of 150 people on board, fifty died in the disaster. (The ship was salvaged in 1961 and now sits, beautifully preserved, in the Vasa Museum in Stockholm.)

Figure 14.1: The Vasa founders in Stockholm harbour on her maiden voyage

Analysis of the circumstances leading to the loss of Vasa (Borgenstam & Sandström, 1995; Ohrelius, 1962) revealed several weaknesses in organisational resilience. Treating the Vasa capsize as a new-product disaster, Kessler et al. (2001) identified contributing causal factors, including:

- Obsession with speed (productivity).
- Top management meddling.

At the time of the Vasa construction, Sweden's military campaigns were under threat. In 1625, ten naval ships had been lost, resulting in a need to accelerate the development of the Vasa. Moreover, their arch-rivals the Danes, were building a super-sized warship, and on learning of this, the Swedish King was determined to have a ship with as many heavy guns as possible, and added a second gundeck and additional cannons to the Vasa's specifications. He had approved the Vasa's dimensions and was keen to have her completed rapidly to strengthen his naval power, but continuously requested design changes in ornamentation and armature. According to Kessler et al. (2001, p. 89), "On several occasions the master shipbuilder cautiously tried to dissuade the king, but the king would not listen. Some speculate the extreme vanity of the king was to blame for his overly grandiose goals and over-involvement." Hybbertson, the master shipbuilder responsible for the Vasa, then became ill and died in 1627, and the ship's construction was left under the supervision of another manager. The stability test on the ship in the summer of 1628 had to be halted when she heeled violently in response to thirty men running across her decks. Nevertheless, the Admiral in charge of the test decided not to delay the commissioning of the ship, and less than a month later, she sank. It appears that weaknesses in managerial resilience, especially the failure to withstand the 'drive for production' from the authority of the King, were instrumental in the disaster.

So it is proposed that the resilience of middle managers is a critical component of organisational safety. They should be able to function as a protective buffer between the competing demands of production (usually driven by senior management) against the safety needs of people, plant and environment. In the case of the Vasa, it appeared that this essential managerial resilience was deficient. For instance, the senior shipbuilder who had expert knowledge and therefore some degree of power, had died before the final year of construction. The Admiral running the sea trials could see that the ship was unstable, but did not dare to delay the launch date (Ohrelius, 1962). One cannot be over-critical of their decisions; in 17th Century Europe disregarding the wishes of your King could have fatal consequences, so their reluctance to contradict Gustavus Adolphus was certainly understandable. But 350 years later, many modern accidents have similar hallmarks indicating a loss of managerial resilience: The regal is power now located in senior management boardrooms and government offices. Dangerous practices can be tacitly encouraged by management even though they contradict formal safety policies (Cutler & James, 1996; Nichols, 1997).

Managerial resilience at NASA also failed in the loss of the space shuttle Challenger in 1986, with powerful political and commercial pressures being applied by senior managers to keep the shuttle programme on schedule. On the eve of the fatal launch, when the engineers were concerned about the fragility of the O rings in low temperature and were voting against the launch, one of the engineers who was hesitating was asked to "take off his engineering hat and put on his management hat" (Vaughn, 1996, p. 398). This engineer then voted in favour of the launch. When the launch took place the next morning in low temperature, the O rings failed and the shuttle exploded on take-off. As Vaughn's analysis reveals, the behaviour of the engineer has to be interpreted in the context of the prevailing organisational culture in NASA, drifting towards the acceptance of increasingly risky operations. Less than a decade later, in 2003 when the Columbia shuttle was damaged on take-off and exploded on re-entry, the accident investigation shows a similar pattern of flaws in NASA managers' decision-making in response to schedule pressure (Gehman, 2003). According to Woods (2003, p. 8) this created 'strong incentives to move forward and look askance at potential disruptions to schedule'.

Other industries are not immune to breakdowns in managers' resilience. Ill-advised trade-offs between production and safety in the energy industry regularly feature in accident analyses (Carson, 1981; Hopkins, 2000; Wright, 1986). In the Piper Alpha oil platform disaster, which killed 167 men (Cullen, 1990), there had been continual pressure for production, coupled with cost cutting but little evidence of any resistance to this (Pate-Cornell, 1990). Hopkins (2005) outlines a similar situation in the Australian airforce where planes became more important than people and consequently 400 maintenance workers were poisoned by exposure to toxic chemicals. So what exactly characterises managerial resilience and how can it be measured and trained?

Managerial Resilience

Resilience has become a fashionable concept in the corporate world (Coutu, 2002; Sutcliffe & Vogus, 2003; Jackson & Watkin, 2004) with diagnostic tools available, such as the Resilience Factor Inventory, to measure individual levels of resilience (Reivich & Shatte, 2002). This is typically discussed at the level of the employee coping with workplace demands and based on the notion of resilience to life's adversities. Luthans and Avolio (2003) point out that the application of resilience to leadership has been largely ignored. Their new model of 'authentic leadership' redresses this by incorporating resiliency as a 'positive psychological capacity' – with themes of realism, improvisation and adaptation. This is a general model of leadership using the interpretation of resilience as the capacity to bounce back from adversity, but it does not make any reference to dealing with safety issues.

The essence of managerial resilience in relation to safety was defined above as the ability to deal with conflicts between safety and the primary performance goals of the organisation. The main goal is usually pressure for production: in industrial settings, there can be significant pressure (explicit or tacit) on site managers and supervisors from more senior managers in the organisation (Mattila et al., 1994). Moreover, there can be internalised or intrinsic pressures from the professionalism of the individual manager to complete tasks, meet targets, reduce costs. These combined pressures can influence workforce behaviour in a manner that drives the organisation too close to its risk boundary, as the Vasa, Challenger and other accidents have shown. Thus, part of the middle manager's resilience must be the ability to recognise encroaching danger from antagonistic risk situations and then to make trade-off decisions between business targets and safety considerations. Safety-conscious companies operating in higher risk sectors, such as energy production, do not pretend that these conflicts do not exist. Malcolm Brinded (currently CEO of Shell Europe) has stated, 'Though there are many occasions on which changes to improve business performance also deliver improved safety, we must ensure that everybody is clear that "if there is any conflict between business objectives, safety has priority"' (Brinded, 2000, p. 19).

Three component skills characterise managerial resilience in relation to safety. The first is *Diagnosis* – detecting the signs of operational drift towards a safety boundary. For a manager this means noticing changes in the risk profile of the current situation and recognising that the tolerance limit is about to be (or has been) breached. This requires knowledge of the organisational environment, as well as risk sensitivity: but Hopkins (2005) describes how managers can be risk-blind in spite of direct warnings. Weick has long argued for the importance of mindfulness and sensemaking skills in managers, as well as in operating staff (Weick & Sutcliffe, 2001). This component of managerial resilience is essentially the cognitive skill of situation awareness which encompasses gathering information, making sense of it and anticipating how the present situation may develop (Banbury & Tremblay, 2004; Endsley & Garland, 2002).

The second component is *Decision-making* – having recognised that the risk balance is now unfavourable (or actually dangerous), managers have to select the appropriate action to reduce the diagnosed level of threat to personnel and/or plant safety. These are relevant across work settings. Woods, (this volume, Chapter 2) calls these *trade-off* or *sacrificial* decisions and argues that these were not being taken when NASA's policy of faster, better, cheaper began to produce goal conflicts with safety objectives. Dominguez et al. (2004) studied conversion decisions when surgeons abandon the fast, minimally invasive technique of laparoscopy to switch to a full abdominal incision when the risk balance for the patient is diagnosed as shifting in an adverse direction. In the oil industry, interviews with offshore managers who had faced serious emergencies, showed that their trade-off decisions were key to maintaining the safety of their installation. One described having to go against the advice of his onshore managers (not usually a career enhancing strategy in the oil industry) by dumping a large quantity of expensive drilling mud overboard in order to de-weight and raise the level of his rig in a dangerous storm. He said this action was "too drastic for them but seemed to be the safest thing to do and all on board relaxed when that decision was made" (Flin & Slaven, 1994, p. 21).

In order to accomplish this kind of resilient response, the manager may also require *Assertiveness* skills in order to persuade other personnel (especially more senior) that production has to be halted or costs sacrificed. Dealing with demands from above is a central facet of a middle manager's responsibility for safety. Yet almost the entire literature on managers and industrial safety concentrates on how managers should behave with their subordinates (e.g., Hofmann & Morgeson, 2004). What many supervisors and managers actually find more difficult is how to effectively challenge their boss when they believe that safety may be in conflict with production goals. During the *Piper Alpha* oil platform accident, the two adjacent platforms *Claymore* and *Tartan* were inadvertantly feeding *Piper's* catastrophic fire with oil because they were so reluctant to turn off their own oil production. (Platforms can take several days to return to full production following a shut-down.) According to the Public Inquiry report, (Cullen, 1990) the production supervisor on *Claymore* asked his boss, the offshore installation manger, to close down oil production on six separate occasions without achieving a result.

These three skills characterising managerial resilience (diagnosis, decision-making, assertiveness) are influenced by a manager's underlying attitudes, particularly *Commitment to safety*. Managers' implicit attitudes can be revealed in sharp relief when production and safety goals are clashing and critical decisions must be taken.

Safety Culture and Managerial Resilience

Whether or not managers make sacrificial decisions in favour of safety depends not only on their skills and personal commitment to safety but on the general level of commitment to safety in the managerial ranks of the organisation. This is the essential ingredient of the organisation's safety culture (Flin, 2003; Zohar, 2003) which affects behaviours such as balancing production and safety goals, implementing safety systems, spending on safety. An effective safety culture produces a belief that when safety and production goals conflict, managers will ensure that safety will predominate.

The organisational culture also has to be sufficiently favourable to allow workers and managers to speak up when they are concerned about safety. Staff need to be sure that when they do challenge their boss, order production to be stopped or express concern about risks, that they will not be penalised. Unfortunately this has not always been the case and, certainly in the UK, there are some striking examples of workers having inordinate difficulty in persuading anyone to listen to their concerns, or even being ostracised for expressing them. In the Bristol hospital where two paediatric cardiac surgeons were operating on babies despite unusually high failure rates, Stephen Bolsin, the anaesthetist who endeavoured to raise concern about their unsafe performance, was given no support and it took several years before the scandal was fully revealed (Hammond & Mosley, 2002; Kennedy, 2001). In fact, Bolsin displayed all three resilience skills described above: he recognised the risks, took the appropriate decision that the situation was unsafe and something had to be done, and (at considerable risk to his own career) spoke up. But despite his courage and assertiveness, the culture of the British medical profession did not endorse such behaviour and his efforts were thwarted. By all accounts this resulted in the avoidable deaths of babies treated by the two surgeons before they were finally suspended. In the UK, 'whistle blowing' legislation has now been introduced (Public Interest Disclosure Act 1998) which is designed to protect workers who disclose information about dangers to health and safety in their organisation.

Measuring Managerial Resilience

Three possible techniques to measure managers' safety resilience are discussed, safety climate surveys, upward appraisal and identifying managerial scripts.

Safety Climate

Safety climate questionnaires measure the underlying organisational safety culture (Cox & Flin, 1998) and provide a leading indicator of safety that complements lagging data, such as accident reports. They can be used as one measure of managerial commitment to safety as they record workforce perceptions of the way in which safety is managed and how it is prioritised against other business goals (e.g. production, cost reduction). When safety climate surveys were first conducted in the North Sea oil industry, they showed that a significant percentage of workers were sceptical about management commitment to safety. In one study on six offshore platforms, 44% of workers agreed that *'there is sometimes pressure to put production before safety'* and that *'pointing out breaches of safety instructions can easily be seen as unnecessary hassle'* (Flin et al., 1996). Two years later, the culture had not changed, in a safety climate survey of ten installations, only 46% of workers agreed that *'Management onshore are genuinely concerned about workers' safety'* and just 23% believed that *'nowadays managers are more interested in safety than production'* (Mearns et al., 1997).

Upward Appraisal

In one of the North Sea oil companies, the managing director was so concerned about his offshore workforce's scepticism about his managers' commitment to safety, revealed in their safety climate data, that he commissioned an upward appraisal of the managers' safety attitudes. It was designed to determine how effectively senior managers communicated their safety commitment to their immediate subordinates/direct reports (who in this case were also senior managers). An upward appraisal exercise of this type can also provide a measure of managerial prioritisation of safety against other goals.

There are no standard tools for measuring safety commitment and safety leadership in an upward appraisal exercise, although 360° appraisal is widely used in assessment of business performance. Therefore, a special questionnaire was designed which included sections on safety commitment behaviours, prioritisation of safety, production, cost reduction and reputation, as well as a leadership scale (Yule, Flin & Bryden, under review). For example, one question asked *'if he was to fail on one of the following business drivers which one would concern him most (production, reputation, safety, cost reduction)?'* The questionnaire was given to 70 directors and senior managers, including directors from their major contracting companies. Each manager completed the questionnaire describing his own safety attitudes, behaviours and leadership style. He also gave a mirror-version of the questionnaire to five direct reports and asked them to rate him in the same way.

The senior managers attended a one-day safety workshop during which each was given a personal report describing his self-perception of safety commitment, contrasted against the view of his subordinates (shown as average and range data). Aggregate results were prepared for the group and presented, resulting in a frank discussion of whether senior managers were successfully communicating consistent messages about their safety commitment. The exercise produced a very positive response from the managers involved, with subsequent evidence of managers taking action to change their behaviour in relation to safety management. Undertaking an upward appraisal survey of this type to assess senior managers' behaviours, such as prioritisation of safety or their response when safety concerns are raised, is another indicator of whether the prevailing culture is likely to support resilient behaviours in managers and supervisors.

Managerial Scripts

Zohar & Luria (2004) have developed an interesting technique based on behavioural scripts for identifying the extent to which managers prioritise production over safety, and the influencing conditions. They explain that, '*From the perspective of employee safety as a focal goal, the primary attribute of managerial action patterns is the priority of safety relative to that of competing goals, such as production or costs*' (Zohar & Luria, 2004, p. 323). They gave platoon leaders from the Israeli army a set of eight infantry scenarios involving competing safety and mission demands and asked them to say whether the mission should continue or be aborted. They used the responses to measure the leaders' safety priority and found that this interacted with leadership style to influence platoon injury rates. This method could easily be adapted for use in other organisational settings to reveal managerial risk awareness and decision-making in relation to safety versus production goals.

Training Managerial Resilience

There are organisations (e.g. the emergency services, military) whose incident commanders are trained to be resilient, i.e., to develop the diagnostic, decision-making and assertiveness skills outlined above (Flin, 1996). Although their risk acceptance thresholds tend to be much higher than those of industrial managers, they too have to make difficult trade-off decisions, such as sacrificing the few to save the many, or halting an operation because the risks exceed the rewards. As Bigley & Roberts (2001) have argued, there may be much to learn for the business world by developing a better understanding of the organisational systems used by the emergency services. The skills of incident commanders translate into resilience competencies that managers could acquire to enhance organisational safety. For example, dynamic decision-making used by fire officers could be applied to enhance risk assessment skills in industry (Tissington & Flin, in press), and there are techniques for training situation awareness and decision-making (Crichton et al., 2002).

Assertiveness training can include influencing skills (Yukl & Falbe, 1992), as well as challenging behaviours and these are already taught in some industrial settings, most notably aviation. Following a series of aircraft accidents, where co-pilots realised that there was a serious risk but failed to challenge the captain's actions, the need for pilots to receive assertiveness training was identified. This gives pilots specific advice and practice on the communication methods to make their challenge in an assertive manner without being passive or aggressive. One large airline considered changing the term co-pilot to 'monitoring and challenging pilot' to emphasise the safety-critical nature of this role.

All three of these resilience skills – risk sensitivity, decision-making, assertiveness – are trained in operational staff in high risk organisations under the banner of Crew Resource Management (Flin et al., in preparation; Wiener et al. 1993). Of course, any organisation that provides this type of training is already moving to a safety culture that has acknowledged the need for such behaviours and is going to accept and reward them as essential for the maintenance of safety.

Conclusion

Organisational resilience has emerged as a conceptual antidote to the traditional pathological approach to safety and disaster research. It reflects the *Zeitgeist* in positive psychology to develop an understanding of human strengths (Seligman, 2003) rather than the prevailing fascination for cataloguing frailties. So does the resilience approach offer us a paradigm shift or is it just a more positive repackaging of a century of ideas on industrial safety? I suspect the latter but it is encouraging a re-analysis of pervasive, intractable safety problems and the focus on positive attributes provides more constructive advice for managers. This chapter has examined one component of organisational resilience, namely managerial resilience – defined as the ability to deal with conflicts between safety and performance goals. As Quinn (1988) has pointed out, leadership is by its very nature inherently paradoxical and the most resilient managers are those that can recognise and respond appropriately to the paradoxes of risk.

Chapter 15

Learning How to Create Resilience in Business Systems

Gunilla Sundström
Erik Hollnagel

In the context of the present discussion, we adopt a simple working definition of *resilience* as an organisation's ability to adjust successfully to the compounded impact of internal and external events over a significant time period. Within the domain of business systems we further define *success* as the organisation's ability to maintain economic viability as defined by the economic markets. We finally define *time period* as the average time period any firm is part of a critical economic indicator list such as the DOW Jones Industrial market index.[1]

To learn about organisational resilience, we adopt a two-pronged approach:

- We first identify implications of adopting a systemic approach to organisations, or, business systems. Hereby we primarily build on work by Ludwig von Bertalanffy as well as more recent work by Senge (1990), Sterman (2000) and Hollnagel (2004).
- We then use a catastrophic organisational failure to illustrate the consequences of lack of resilience. The case in point is the downfall of Barings plc in 1995. The analysis identifies actions that potentially might have helped Barings plc to reduce the combined

---

[1] Of the original companies forming the DOW Jones only one (i.e., General Electric) made it to the 100th anniversary of the index.

catastrophic impact of the actions of a single individual and market events.

The System View: Implications for Business Systems

The originator of general system theory was Ludwig von Bertalanffy, an Austrian biologist devoted to finding principles that could be applied across various scientific disciplines. The book "Perspectives on General System Theory" published in 1975, three years after von Bertalanffy's passing, provides an overview of some of the most central concepts of the proposed system approach. The aim of the following subsections is to apply some of the most important concepts suggested by von Bertalanffy to business systems.

The Organism and System Concepts

Two critical concepts in von Bertalanffy's thinking are the concept of *organism* and the closely related concept of a *system*. von Bertalanffy introduced the organism concept to create a contrast to a machine theoretic approach to biological entities (cf. von Bertalanffy, 1952, pp. 9). He stated that

> ... organic processes are determined by the mutual interaction of the conditions present in the total system, by a *dynamic* order as we may call it. This is the basis of organic regulability ... Organisms *are not* machines, but they can to a certain extent *become* machines, congeal into machines. Never completely, however for a thoroughly mechanised organism would be incapable of regulation following disturbances, or of reacting to the incessantly changing conditions of the outside world. (von Bertalanffy, 1952, p. 17)

Hence, the organismic concept as developed by von Bertalanffy assumes that resilience is one property that organismic processes and systems have. The term organism (or organised entity) later morphed into the notion of a system as the program of systems theory emerged (e.g., von Bertalanffy, 1975, p. 152).

General system theory defines a system as "... a complex of elements standing in [dynamic] interaction" (von Bertalanffy, 1952, p. 199). The basic focus of scientific investigation is to formulate general principles that guide this dynamic interaction. The type of system that von Bertalanffy focused most attention upon was an open system, i.e., a system that constantly exchanges material – matter and energy – with its environment. In contrast to that, a closed system is defined as one that does not exchange matter or energy with the surroundings, or even as a system that is cut off from its environment and does not interact with it. A closed system must, of course, in practice be able to exchange information with the environment, since without such an exchange it would be impossible to know what happened inside it!

The fundamental research questions emerging from von Bertalanffy's perspective include understanding the underlying principles of system organisation and order, wholeness, and self-regulation. A business firm is clearly an *open* system that exchanges material with its environment. In fact, the open systems view of business firms lead to approaches focused on demonstrating how control mechanisms can be leveraged to reach defined objectives while viewing organisations as organisms (e.g., Morgan, 1986; Lawrence & Dyer, 1983).

The following definition of a business system will be used in the present work: A business system is defined as a set of elements that interact among themselves and with their environments. The overall system is focused on achieving shareholder value, profitability and customer equity. Particular objects, or wholes, within the system are defined by their interactions within the system as well as with the environment. The organisation of these wholes can be described in multiple ways, for instance as hierarchies, heterarchies or networks. A critical point being that the way wholes self-organise changes over time. Some system properties are not directly observable, the property of resilience being among these. To manage a business system, it is necessary to establish a view of the organisational regularities ('laws') at each system level. General management, policies and control principles need to be based on principles applicable to all types of organised entities. To illustrate this point, we provide two examples in the next section.

Examples of System Thinking

The two examples presented in the following illustrate how concepts developed by general system theory have been used to create principles that are potentially applicable across all levels of a particular business system. The first section describes an approach that supports the identification of generic control elements. The second section provides a generic view of types of control behaviour – the assumption being that the behaviour of organised entities, or systems is purposeful, i.e., the system if focused on achieving particular goals. In the case of business systems, we assume that these goals include shareholder value, profitability and customer equity.

Identifying Management Control Entities

A major contribution of control theory and systems engineering (e.g., Sheridan, 1992) has been the introduction of key concepts that can be used across many types of systems to describe system component types including control system components. Adopting a control theoretic and system engineering flavoured approach, a business system can be described as a state machine characterised by a set of variables and parameters. Business goals are associated with desired state values of the business system and the state machine describes how the various transitions among states can take place. For example, defining the desired state of profitable operation in terms of profitability (e.g., that profit margins need to be higher than 30%) and shareholder value, would be a first step towards articulating business goals. The management process of a business system is defined as the control system whose role it is to ensure that the business system meets or exceeds performance objectives. The behaviour of the control system is driven by the defined business goals, such as profitability, shareholder value and customer equity. The business actuators are the means used to change one or more state variables. Sensors provide information about the state variables, either by direct measurement or by some method of estimation. This control system's view of a complex networked business is illustrated in Figure 15.1, adapted from Sundström & Deacon (2002).

The ovals signify that an entity appears as a whole to other entities while the different types of lines indicate that the interactions among entities have different attributes. The basic management philosophy is uniform across the wholes, i.e., based on collecting data and from that deriving metrics and/or estimates to be used by a management control function. The control function takes actions that are designed to impact the behaviour of individual and/or groups of wholes using various types of actuators.

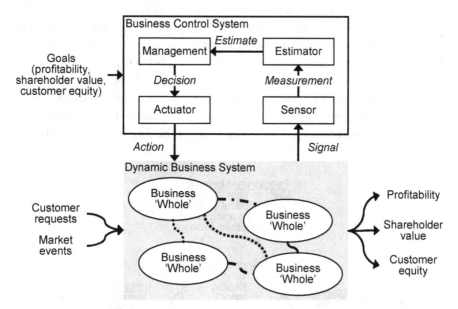

Figure 15.1: A Control theoretic perspective on business systems

Broadly speaking, control functions, i.e., management actions at different dynamic business system levels, are geared towards furthering behaviours that lead to a healthy state of an individual whole, collection of wholes, and/or the overall business system. This healthy state is defined by objectives defined for the whole. State variables are used both to track progress towards these goals (i.e., leading indicators) and to establish whether goals have been achieved (i.e., lagging indicators). Of particular importance are leading measures related to the key resources used by business systems, i.e., financial, human and (technology) systems resources.

For business systems in general, and those focused on financial services in particular, at least three distinct states can be defined as illustrated in Figure 15.2.

A business system can be in a *healthy* state, i.e., a state in which business goals are met and risks are understood and accepted. Various types of behaviours can cause either an individual whole or the complete business state to transition into an unhealthy state. In such an *unhealthy* state the business goals are not met and/or the risk of incurring losses is unacceptably high. Finally, the system can move to a *catastrophic* state, where either one or more individual wholes or the overall system is lost or destroyed. The probability that the overall system transitions into an unhealthy state increases if the behaviour of the wholes creates outcomes that are in conflict with the overall goal states for the business system. The more wholes adopt behaviours that do not consider the overall system goals, the more likely these behaviours will bring about an overall negative impact and, as a result, the overall system might transition from a healthy to a catastrophic state.

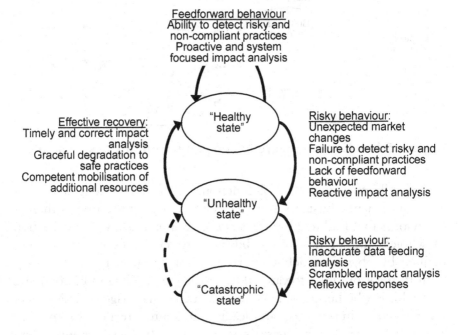

Figure 15.2: Three key business system states transition behaviours

Several interesting questions and implications emerge as a result of viewing business systems as open dynamic systems: (1) Is it possible to identify events that indicate when a system is losing control, hence is in the process of changing from a healthy to an unhealthy state? (2) Is it possible to monitor business systems' operational performance? (3) Is it possible to distinguish healthy system changes from unhealthy changes? All three of these questions should in fact drive control/management behaviour. In the next section, we look at various modes of control/management behaviour.

Models of Control Behaviour

While different domains have different characteristics, a common feature for all domains, including the domain of business systems, is the need to be in control. The successful accomplishment of any kind of activity – from running a nuclear power plant over financial trading to driving to work and cooking *spaghetti carbonara* – requires that the acting entity or agent (manager, team leader, operator, trader, driver, or chef) can maintain control of the process in question.

An essential part of control is planning and forecasting what to do within the system's short-term time horizon. (Planning for the longer term is, of course, also important, but the uncertainty of outcomes is usually so high that only a limited effort in that direction can be justified.) This planning is influenced by the context, by knowledge or experience of dependencies between actions, and by expectations about how the situation is going to develop – in particular about which resources are and will be available. The outcome can be a more or less orderly series of activities, where the orderliness or regularity of performance reflects the degree of control. This can be described in more formal terms by referring to four characteristic control modes, called strategic, tactical, opportunistic and scrambled (Hollnagel, 1993b; Hollnagel & Woods, 2005).

In the scrambled control mode, the choice of the next action is basically irrational or random. This is typically the case when the situation assessment is deficient or paralysed and there accordingly is little or no alignment between the situation and the actions. In the opportunistic control mode, the salient features of the current context determine the next action but planning or anticipation are limited, for instance because there is limited time available. The resulting choice of actions is often inefficient, leading to many useless attempts being made. The tactical control mode corresponds to situations where performance more or less follows a known and articulated procedure or rule. The time horizon goes beyond the dominant needs of the present, but planning is of limited scope or range and the needs taken into account may sometimes be *ad hoc*. Finally, in the strategic control mode, the time horizon is wider and looks ahead at higher-level goals. At this level, dependencies between tasks and the interaction between multiple goals is also taken into account.

Adoption of a strategic control mode is facilitated by adopting system thinking – a critical ability for learning organisations according to Senge (1990, p. 14). A learning organisation (or system) is an entity that "continually expands its capacity to create its future" (Senge, 1990, p.14). Senge discusses five disciplines that will lead to organisational learning: personal mastery, team learning, mental models, building shared vision and system thinking. Senge views the fifth discipline, i.e., system thinking, as the discipline that integrates all others. Senge defines the key elements of system thinking as follows: "... seeing interrelationships rather than linear cause-effect chains, and seeing processes of change rather than snapshots" (Senge, 1990, p.73). (For other examples of the importance to adopt a dynamic system view or system thinking, see Sterman, 2000.) To start a process of system thinking, Senge argues that the first step is to gain an understanding of feedback loops. This will eventually establish an ability to identify recurring event patterns that reflect specific system structures, or system archetypes. To use von Bertalanffy's language, the ability to identify recurring wholes (i.e., patterns of interaction among organised entities) will develop and as a consequence this will facilitate adoption of a feedforward-based strategic control mode.

In summary, the general system view has the following implications for business systems:

• A business system is a dynamic open system with wholes organised on multiple levels, e.g., wholes could be described as being organised in a classical hierarchical structure. Each organised whole is defined by its interactions with other entities as well as its environment.

• Business systems will define desired behaviours by defining goals, policies, standards, processes and procedures. The expectation is that defined policies, processes and procedures should help the wholes of the system to stay in a healthy state. The nature of some of these policies, standards, processes and procedures will depend on what the business system is producing to create market value. For example, if a business system produces financial services, its policies will reflect the fact that financial services is a regulated global industry.

- Each organised whole develops a pattern of interactions driven by its defined and/or perceived goals. Obviously, goals associated with organised wholes can be in conflict with each other and/or in conflict with the business system's overall goals, i.e., profitability and shareholder value. As we will see in the Barings plc example, such a goal conflict eventually drove the organisation into a catastrophic failure.

- Organised wholes can adopt different types of control modes or management behaviour. These can be described as strategic, tactical, opportunistic, scrambled or a mixture of the four, depending on the conditions. Using the Barings example, it will become clear that the most appropriate strategy to maintain control of an open dynamic business system is a feedforward based strategy.

- The property of resilience emerges as a result of a system's ability to transition from one state to the next. In fact the property of resilience implies that a system has the ability to maintain a healthy state over time despite the fact that it (or its wholes) may be subjected to negative and/or destructive events. A key pre-requisite for a system's ability to maintain a healthy state, or to transition from an unhealthy to a healthy state, is to re-organise/re-adjust system boundaries and/or re-align/change both the scope and the types of business controls used as part of the business control system.

Figure 15.3 provides an overview of the concepts used to describe business systems. Three separable business systems are portrayed in Figure 15.3, namely business systems A, B and C. Each system has its own control system, leveraging the control system elements illustrated in Figure 15.1. Thus, each system has defined goals, market drivers and desired output, i.e., measurable shareholder value, profitability and customer equity. The policies, standards, processes and procedures defined by each firm are designed to cover the scope as determined by the system boundaries relative to each system's business control system.

System boundary defined by Firm A's Business Control System

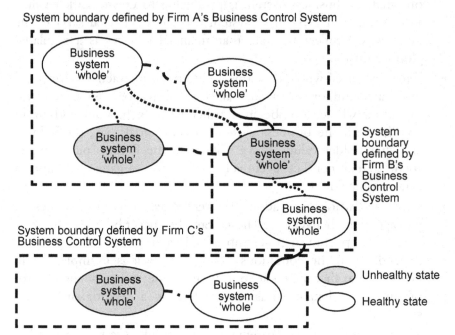

Figure 15.3: A view of business systems and system boundaries

Each business system's whole will develop patterns of interactions to reach its objectives. A business whole that reaches its objectives using behaviours in accordance with policies, processes, standards and procedures is in a healthy state. A key enabler for any business whole to reach its objectives is to be able to predict the impact of its behaviour. As discussed above, the ability to predict impact of behaviour is maximised using a strategic feedforward control approach. As an example, firm B's business control system views a part of firm A as part of its business control scope. As a result, firm B's control approach can proactively adopt a strategic view of the impact of the business whole belonging to firm A. If the reverse is not true, then a likely result is that firm A has a limited ability to change the business wholes' unhealthy state to a healthy state if the state is caused by the interaction with B's business whole. Over time, this situation can have a negative impact on both A and B, or, it can lead to a re-design of the control system scope of firm A to better match firm B's approach. As a result, both firms might end up being more resilient to potential negative impact resulting from the two firms' interactions.

As we will see in the Barings example, the lack of explicit design of the business control system coupled with the lack of viewing Barings as a dynamic system eventually led to the catastrophic events that destroyed the firm.

The Barings plc Case

Barings plc was a 233 year old British financial institution of high reputation, which proudly counted the Queen as a client. Financial markets were therefore stunned when this institution had to declare a state of bankruptcy in February 1995. The reason was a staggering US$1.3 billion loss caused by the trading practices of a single person, Nick Leeson (e.g., Kurpianov, 1995; Reason, 1997). This loss was, however, not due to the actions of an individual trader alone but also owed much to the failure of Barings' global management team. Figure 15.4, adapted from Sundström & Hollnagel (2004), provides an overview of the key components of the dynamic system of which Nick Leeson was a part.

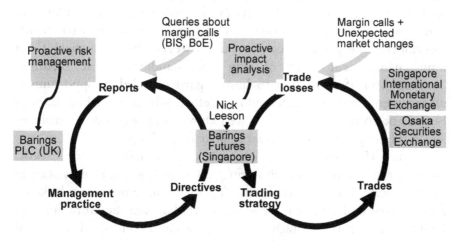

Figure 15.4: Dynamic system view of the Nick Leeson scenario at Barings Securities

The situation can be described as two linked control loops – although neither of them was in full control of the situation. In one, the right-hand part of Figure 15.4, was the trader (Nick Leeson), who saw his trading strategy fail. The outcome of the trades was far from expected (also because of market unpredictability), and he therefore had to resort to a more risky but less well thought out strategy in the hope of covering the losses. In order to hide the losses, Leeson removed his error account from the daily trading, position and price reports to London. As a result any type of proactive impact analysis by the second control loop became impossible. In addition to problems with the market, Nick Leeson also faced requests from the Barings' home office, altogether leading to a situation of fire-fighting and opportunistic – or even scrambled – behaviour rather that tactical thinking.

In the second control loop, the left-hand part of Figure 15.4, was the company, Barings plc, which continuously received reports of stellar profits; in fact profits that were quite unreasonable given the officially communicated securities trading strategy. However, the management's lack of experience and concomitant lack of feedforward strategy, and their inability to truly understand the behaviour of Barings Securities business coupled with the perception of Leeson as a star performer, led to a situation in which the team did not take proper action. During this time period, Barings plc (UK) was itself under pressure to provide information about the (excessive) margin calls to the Bank of England as well as the Bank of International Settlements. (This therefore constitutes a third control loop that had an effect on the two already mentioned. The present analysis will, however, disregard that.) Over time, management's lack of true understanding of the securities business (and of derivatives trading in particular) led to opportunistic management control, which could not effectively monitor how Barings Futures performed. The result is well known: in both cases control was lost and Barings plc had to declare a state of bankruptcy.

Figure 15.5 describes this situation using a state diagram. Clearly, some of the system wholes in Barings plc were not in a healthy state in the first place. From the beginning (t_1), the management team lacked a fundamental understanding of securities trading and therefore failed to establish appropriate management strategies and controls. The lack of understanding also meant that excessive revenue was not seen as a problem. Even as the situation deteriorated (t_2), management at Barings plc ignored early warnings in their balance sheet indicating that they provided excessive funds to run Barings Futures. Instead they approved excessive margin calls, failed to separate sales and reconciliation, and failed to ask for critical data. In the last stage, (t$_3$), the management team failed to notice that the losses incurred exceeded the available capital, but continued to provide excessive funding to run Barings Futures. Despite multiple requests from key regulatory bodies, cf. above, their reporting was not transparent.

Conditions were not much better at Barings Futures in Singapore. The trader, Nick Leeson, was new in his position and lacked experience. He was furthermore in control of both sales and reconciliation, contrary to policies. As the situation grew worse, (t$_2$) and (t$_3$), he continued with excessive margin calls, tried to hide the reconciliation and removed key data from reports. Thus even if the Barings plc management team had known what to look for, they would have had difficulties in finding it.

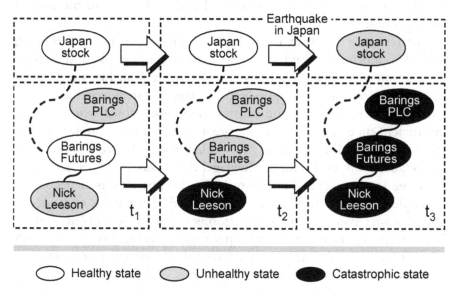

○ Healthy state   ◯ Unhealthy state   ● Catastrophic state

Figure 15.5: A state transition view of the Barings plc scenario

A key learning from this scenario is that a business system can only be resilient if the management team is able to use a feedforward based management strategy resulting in the appropriate design of business controls. Examples of such critical business controls are an appropriate flow of reporting information and clarity around roles and responsibilities. Both of these obviously impact how the various business system wholes interact with each other.

An organisation is resilient if it is able successfully to adjust to the compounded impact of internal and external events over a significant time period. Barings plc was clearly unable to do so, and given the above analysis its collapse was not just due to the specific circumstances of Nick Leeson's trading but rather to a general degradation of the barriers and behaviours that together made up its control systems. In this case each of the two business control systems, Barings Futures in Singapore and Barings plc in London, failed in adjusting effectively to exogenous variability. For Barings Futures, the exogenous variability came from the markets; for Barings plc the variability came from Nick Leeson's trading practices. Both resorted to feedback-driven control and therefore worked in a basically reactive mode. This is bound to fail if the feedback is incomplete or delayed, as in the case of Barings plc, or if the response strategy is short-sighted and shallow, as in the case of Barings Futures.

In terms of the four control modes mentioned above (strategic, tactical, opportunistic and scrambled), both Barings plc and Barings Futures operated in an opportunistic or even scrambled mode. There is in general a strong correspondence between system conditions and control modes, such that the control mode goes down when there is insufficient time and/or when the situation becomes unpredictable. (A more detailed discussion of the relationship between time and predictability can be found in Hollnagel & Woods, 2005.) Referring to the three states described in Figure 15.5, a healthy state is characterised by strategic and tactical control, an unhealthy state by tactical and opportunistic control, and a catastrophic state by scrambled control. While recovery from a state with opportunistic control is possible with a bit of luck, i.e., if conditions are favourable, recovery from a state of scrambled control is unlikely to take place.

What would have made Barings more Resilient?

The first step Barings' management team should have taken is to view their trading business from a dynamic open system perspective. Figure 15.6 provides a simple view of the trading system as a dynamic system.

If Barings had used a proactive impact analysis and risk management approach, and if it proactively had monitored its trading operations to check for unusual performance variability, then Leeson's trade outcomes should have been anticipated – by Barings plc, if not by Leeson himself. The proactive analyses should have enabled Barings to anticipate patterns and therefore to be able to proactively take actions (feedforward-driven) rather than just reacting to events (feedback-driven). In such a feedforward-driven mode, system control behaviour is triggered by a model of derivatives trading and likely events. In a feedback-driven mode, system control behaviour is a reaction to undesirable events (e.g., Leeson's frequent margin calls). If trade outcomes had been predicted, including expected profits, discrepancies would have been noted in time and Barings would have remained in control. However, the Barings' management team was in a reactive mode; the team furthermore had no access to the appropriate data nor did they take any action to ensure that Leeson was appropriately supervised. As a consequence, the Baring business system was highly unstable and as such vulnerable to disturbances from the market. This combination of reactive decision-making in response to Leeson's behaviour and providing regulatory entities (i.e., Bank of England) with poor information, over time ended in a complete failure when unexpected events affected the market in a negative direction.

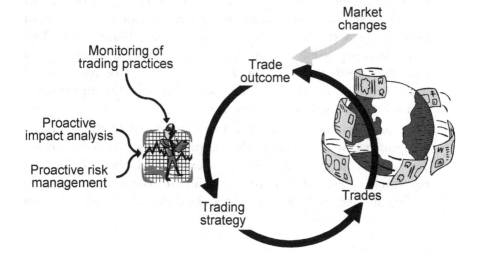

Figure 15.6: Dynamic system view of trading operations

If Barings' management team had adopted a feedforward management strategy, they would quickly have known that the revenue reported by Nick Leeson would have been impossible given the trading strategy that Leeson purportedly was using. However, adopting a feedforward strategy assumes that the management team has sufficient knowledge about the nature of the securities business. In Barings plc this was however not the case, most of the management team had a traditional banking background and lacked experience with the securities industry.

Assuming that Barings' management team had adopted a dynamic system view of their business, the team could have proactively established a feedforward strategy with the following elements:

- Clearly defined system goals, including the level of risk acceptable to both the local Asia-based and London-based management teams. These system goals should have been designed seamlessly to intertwine business and risk goals. For example, a part of the risk management strategy could have been tightly to control the size of the margin calls approved before providing Leeson with funds for his trading activities.

- A continuous monitoring of state variables, i.e., those variables that could change the state of Barings from healthy to unhealthy. In the Barings case the management team did not seem aware of the fact that they lacked accurate data to identify the potential impact of Leeson's activities on the whole system. i.e., Barings plc. This situation was possibly due to management's lack of knowledge of what they needed to look for. In fact, a very simple report could have been used to match up the losses made and the capital made available to Baring to cover these losses.

- Continuous monitoring of state variables related to the three key components of operational risk, i.e., people, systems and process.[2] This should have included close supervision of Nick Leeson, the reporting systems that were used and of course the associated trading and reconciliation processes. As it turned out, Nick Leeson

---

[2] See http://www.bis.org for information about the Basel II accord and operation risk.

was not supervised and the required data to monitor his activities were not available to the London-based management team. Finally, trading and reconciliation processes were not properly separated, i.e., Nick Leeson was in control of both.

Figure 15.7 provides a high-level view of the simple management framework that Barings could have leveraged proactively to manage risk and the impact of the Barings Securities business on the overall Barings business system. The Barings example also clearly demonstrates the importance of identifying unexpected (i.e., unlikely) events that have the ability to drastically impact the overall state of the business system.

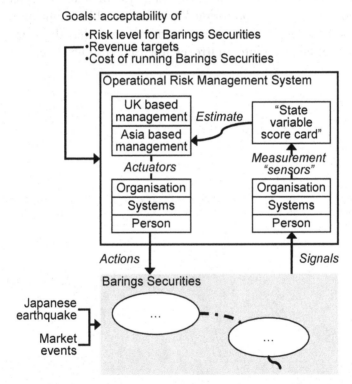

Figure 15.7: A simple control engineering style approach to support a feedforward driven proactive risk management strategy

Concluding Remarks

An organisation's ability to survive depends on the extent to which it is able to adjust to the compounded impact of internal and external events over a significant time period. A critical success factor for any organisation is therefore finding ways to learn how to facilitate the emergence of resilience. In the present chapter, we suggested that viewing organisations as dynamic open systems, adopting an explicit view of business control systems and being aware of the impact of various types of system control behaviours could enable an organisation to avoid transitioning into a catastrophic state. Our analysis of the Barings plc case illustrated that articulating system goals, acceptable risk levels and a focus on monitoring key state variables could potentially have helped Barings' management team to detect that the firm was in the process of entering into an irreversible catastrophic state.

Chapter 16

Optimum System Safety and Optimum System Resilience: Agonistic or Antagonistic Concepts?

René Amalberti

Introduction: Why are Human Activities Sometimes Unsafe?

It is a simple fact of life that the variety of human activities at work corresponds to a wide range of safety levels.

For example, considering only the medical domain, we may observe on the one hand that audacious grafts have a risk of fatal adverse event (due to unsecured innovative protocol, infection, or graft rejection) greater than one per ten cases (10^{-1}) and that surgeries have an average risk of adverse event close to one per thousand (10^{-3}), while on the other hand fatal adverse events for blood transfusion or anaesthesia of young pregnant woman at delivery phase are much below 1 per million cases (see Amalberti, 2001, for a detailed review). Note that we are not, strictly, speaking here of differences that are due to the inherent severity of illnesses, but merely on differences due to the rate of errors and adverse events.

Another example is flying. It is usually said that aviation is an 'ultra safe' system. However, a thorough analysis of statistics shows that only commercial fixed wing scheduled flights are ultra safe (meaning that the risk is 10^{-6} or lower). Chartered flights are about one order of magnitude below (10^{-5}), helicopters, scheduled flights and business aviation are about two orders of magnitude below (10^{-4}), and some aeronautical leisure activities are at least three orders of magnitude below (10^{-3}), cf. Figures 16.1 and 16.2.

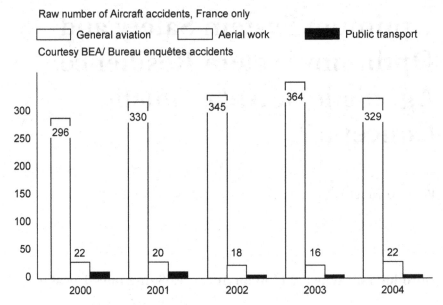

Figure 16.1: Raw number of accidents in 2004 in France. General aviation accident rate is about two orders of magnitude of commercial aviation

In both cases, neither the lack of safety solutions nor the opportunity to learn about their use may explain the differences. Most safety tools are well known by stakeholders, available 'on the shelf', and implemented in daily activities. These tools comprise safety audits and evaluation techniques, a combination of mandatory and voluntary reporting systems, enforced recommendations, protocols, and rules, and significant changes in the governance of systems and in corporate safety cultures.

This slightly bizarre situation raises the central question debated in this chapter: If safety tools in practice are available for everyone, then differences of safety levels among industrial activities must be due to a precisely balanced approach for obtaining a maximum resilience effect, rather than to the ignorance of end users.

One element in support of that interpretation is that the relative ranking of safety in human activities at work has remained remarkably stable for decades. Safety is only slowly improving in parallel for most of these activities, available figures showing an improvement of an order of magnitude during the last two decades for most human activities.

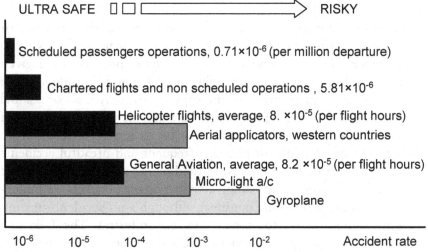

Courtesy :
Boeing statistical summary of commercial jet airplane accidents, 1959-2000
www.helicoptersonly.com/sayagain_danger.html (accessed May 25, 2005)
Bureau des enquêtes accidents France, www.bea-fr.org
National Transportation Safety Board, www.ntsb.gov

Figure 16.2: Accident rate in several types of aviation

Only a very central set of human activities have seen their safety improving faster than average, and when considering these activities, e.g., medicine, the reason is usually a severe threat to the economy and the resilience of the system rather than because safety as such was seen as not being good enough.

For example, the impressive effort in the US made in patient safety since 1998 was directed by the growing medical insurance crisis and not by any spontaneous willingness of the medical community to improve safety (Kohn et al., 1999). Consistent with this, once a new level of resilience has been reached the system re-stabilises the balance and does not continue to improve. This levelling off can be seen in medicine. After an initial burst of enthusiasm and an impressive series of actions that led to a number of objective improvements, the second stage of the 'safety rocket' is failing because very few medical managers want to continue. It is as if a new resilience should have been reached, with a new set of rules, a new economic balance, and therefore, no need to go further (Amalberti et al., 2005; Amalberti & Hourlier, 2005).

To sum up, this paper suggests two supplementary interpretations:

- *Existence of various types of resilience.* That there are various types of resilience corresponding to classes of human activities. The resilience should be understood as a stable property of a given professional system, i.e., the ability to make good business in the present and near future with the capacity to survive an occasional crisis without changing the main characteristic of the business. Of course, the safety achievement may be extremely different from one type of resilience to another: frequent and dreadful accidents may be tolerated and even disregarded by the rest of the profession (ultra-light biological spreading activities against crickets, or mining accidents in third world countries), although a no-accident situation is the unique hope for others (nuclear industry). The following section will propose a tentative classification of the types of resilience.

- *Incentives to change.* Jumping from one stage of resilience to another, adopting more demanding safety logic is not a natural and spontaneous phenomenon. A professional system changes its level of resilience only when it cannot continue making effective business with the present level of resilience. It is hypothesised in this chapter that there are two major types of incentives to change. One is the external incentives due to the emergence of an economic or/and political crisis. The crisis can follow a 'big accident', a sudden revelation by media of a severe threat or a scandal (blood transfusion in France, unsafe electricity system in northern US, medical insurances in all other Western countries), or a political threat (patient safety in UK and re-election of Tony Blair). Another is internal and relates to the age of the system. Indeed, systems have a life cycle similar to humans. When aging, the systems are 'freezing', not necessarily at the highest level of safety, but with the feeling that there is no more innovation and resources to expect, and that a new paradigm is required to make the business effective in the future. The following sections detail these two incentives to change, external and internal.

Examples: Concorde and the Fireworks Industry

For example, when the Concorde aeroplane was designed in the late 1960s it was in perfect compliance with regulations existing at the time. Since it was impossible to make any retrofit later in the life cycle of the aircraft (because of the very few aeroplanes existing in the series and the unaffordable cost of making retrofits), the Concorde progressively became in breach of the improvements in Civil Aviation regulations. At the end of its lifetime, in the late 1990s, this airplane was unable to comply with many of the regulations. For example, adhering to the new rule of flying 250 kts maximum below flight level 100 was not possible (as there would not be enough fuel to cross the Atlantic if complying). It was also unable to carry passengers' heavy luggage (choices had to be made between removing some fuel, some seats, or some luggage) and this limitation was totally in contradiction with the increased use of Concorde for prestige chartered flights. In both cases, solutions were found to resist and adapt to growing contradictions, redesigning the operative instructions to be on the borderline of regulations but compatible with the desirable outcome. To sum up, for the continuation of Concorde flight operations, adopting all of the new safety standards of civil aviation without compromising would have resulted in grounding the aeroplane. The resilience of Concorde, playing with boundaries of legality, was so good that Concorde survived for three decades with borderline tolerated conditions of use [BTCUs] before being definitively grounded (see Polet et al., 2003, for a theory on BTCUs). And even after the Paris crash on July, 25th, 2000, that killed 111 persons, Concorde continued to flight for two more years despite evidences that it should have been definitively grounded. Again this strange result can be explained by the superb resilience of the Concorde operations in British Airways and Air France (of course, in that case the resilience unfortunately does not apply to the Concorde having crashed, but applies to the survival of global operations). Concorde was more than an aeroplane, it was a European flag. As such, it was considered that the priority of resilience should be put on the continuation of operations all around the world and not on ultimate safety.

The story of Concorde is just one example among many in the industry. For another, consider the fireworks industry. The average safety standards of this type of industry remains two orders of magnitude below the safety of advanced chemical industries. However, when looking carefully at their logic, they deal with extremely unstable conditions of work. The job is based on a difficult crafting and innovative savoir-faire to assemble a mix of chemical products and powders. Innovation is required to remain competitive and products (made for recreational use only) must be cheap enough to be affordable to individuals and municipal corporations. All of these conditions make the fireworks industry a niche of small companies with low salaries. Finally, a common way of resilience to ensure survival is to adapt first to the risk, save money and refuse drastic safety solutions. One modality of adaptation has been illustrated by the national inquiry on the chemical industry in France conducted after the AZF explosion in Toulouse in September 2001. The report showed how firework factories leave big cities in favour of small and isolated villages where a significant number of inhabitants tolerate the risk because they are working in the firework factory. In that case, resilience (i.e., industry survival) has reached a safety deadlock. Accepting to improve safety would kill this industry. That is exactly what its representatives advocated to the commission, threatening to leave France if the law had to be respected. (Note that safety requirements already existed, and the problem was, in that case, either to enforce them or to continue tolerating borderline compliance.)

Mapping the Types of Resilience

The analysis of the various human activities at work and their safety-related policies lead us to consider four classes of system resiliencies (Table 16.1).

Ultra-Performing Systems

A first category of human activities is based on the constant search and expression of maximum performance. These activities are typically crafting activities made by individuals or small teams (extreme sports, audacious grafts in medicine, etc.). They are associated with limited fatalities when failing. Moreover, the risk of failure, including fatal failure, is inherent to the activity, accepted as such to permit the maximum performance. Safety is a concern only at the individual level.

People experiencing accidents are losers; people surviving have a maximum of rewards and are winners.

Table 16.1: Types of systems and resilience

	Safety level (Order of magnitude)			
	Ultra performing system (Safety < 10^{-3})	Egoistic system (Safety < 10^{-5})	Collective expectation (Safety < 10^{-6})	Ultra-safe systems
Types	Crafting industry, Mountaineering, Extreme sports, Transplant surgery	Drivers using the road facilities, or patients choosing their doctors	Food industry, banks, services, incl. medical, as part of a standard accompanying package set	High risk complex systems: Transportation, energy
Model of success	Outstanding performance. Constant search and expression of maximum performance	Individual satisfaction. The choice of service fully depends on customer decision	Collective satisfaction. Customers are not making a direct choice of these services	No accidents. One accident anywhere means the end of business
Model of failure	Low competence	Poor team work. Unstable quality and delivery. Fatalities here and there. Individual victims sue individual workers.	Poor top organisation. Accidents possible. Victims form a group and sue local social entities and local politicians.	Complacency. Large accident is the rule. Victims form a group and may sue system as a whole
Criteria for resilience	Training competitiveness	Quality and control procedures	Transparency, HRO, management regulations	Show compliance, accept supervision, ready for the 'big one'
Who is in charge of organising resilience?	Everyone and no one	Quality managers. Business risk control department	Safety managers	Safety managers scrutinised by international & government agencies

Egoistic Systems

A second category of human activities corresponds to a type of resilience based on the apparent contradiction between global governance and local anarchy. The system is like a 'governed market' open to 'individual customers'. This is for instance the case for drivers using the road facilities or patients choosing their doctors. The governance of the system is quite active via a series of co-ordinations, influencing of roles, and safety requirements. However, the job is done at the individual level with a limited vision of the whole, and logically the losses are also at the individual level with fatalities here and there.

One of the main characteristics of this system is that the choice is made at the users' level, directly beneficial to him or her, and therefore exposes most of the safety loopholes. For example, the patient may choose his/her doctor, or even decide on proposed medical strategies with pros and cons. General requirements, such as quality procedures, are sold as additional criteria that may influence the choice of the customer. Since accident causes mix the choice of end users and of professionals, they tend to be considered as isolated problems with little or no consequence for the rest of the activity. The usual vision is that losers are poor workers or unlucky customers. There are very few inquiries, and most complaints are based on individual customers suing individual professionals. Systems of that type are stable and do not seek better safety, at least as long as the economic relationships remain encapsulated in these inter-individual professional exchanges. However, should the choice no longer be the prerogative of individuals, the whole system is immediately put at risk and therefore moves to a new balance, adopting the rules of resilience of the next safety level. Regarding safety tools used at that stage, the priority is to *standardise* people (competence), work (procedure) and technology (ergonomics) with recommended procedures in design (ergonomics) and operations (guidelines, protocols) as the main generic tools. The tools for standardisation will then move on to more official prescriptions at the next level, and ultimately will be turned into new federal or national laws in the last stage.

Systems of Collective Expectation

The third category of human activities corresponds to public services and large low risk public industries. Here people in the street are not making direct choices. This is typically the case of the food, post-office, energy supply, bank supply, and a series of medical accompanying services (general hygiene conditions, sanitary prevention, biology, etc.).

These services are considered as necessary at the town and region level. Their failures can lead to social chaos, directly engaging the competency and commitment of the local top managers and politicians. This is the reason why resilience at this level consistently incorporates a high public visibility and communication effort on risk management. Safety bureaus with safety officers deliver public reports giving proofs of safety value on a regular basis. To sum up, resilience of that type mainly consists in showing a higher commitment, developing transparency and communication on safety, and managing the media, pushing evolution but not revolutions, and never abandoning the local control of the governance. However, with this evolution of governance, safety strategies benefit from new impulses. A continuous *audit* is required to address and control residual problems. This may require an extensive development of monitoring tools such as in-service experience, reporting systems, sentinel events, morbidity and mortality conferences. It is an occasion to consider safety at a systemic level, enhancing communication and the safety culture at all levels, including the management level. The teaching of non-technical skills (such as Crew Resource Management) becomes a priority in order to get people to work as a team. Macro-ergonomics tend to replace local or micro-ergonomics, and policies tend to replace simple guidelines. The audit procedure can identify a series of recurrent system and human loopholes, which resist safety solutions based on standardisation, education, and ergonomics. All solutions follow an inverted U-curve in efficiency. When the cost-benefit of a solution becomes negative (for example, the ratio of time spent on recurrent education versus effective duty time), the governance of the system has to envisage transitioning to the next step and the present system of resilience is questioned.

This is typically the effect of the new French law on industrial risk prevention (*loi Bachelot-Narquin*, July 2003) voted after the AZF explosion in Toulouse in September 2001. The AZF explosion destabilised the resilience of the whole chemical industry in France, questioning its presence on French territory. However, the final content of the law issued two years after the accident is the result of a series of contradictory and severe debates on the type of action to be conducted at the nation level for reinforcing safety in the chemical industry (see the Fiterman report, 2003). This law produced a very prudent 'baby'. Instead of changing the foundations and recommending the adoption of a radical change/restrictions in technique and operations to make the job safer (considered as a non-affordable effort and a major threat by the chemical industry unions), the law in the end almost only asks for better visibility, communication and information to neighbours on the risk around factories. The communication procedure relies on the creation of 'local citizen consulting committees' with a mixture of industrial and citizen representatives. To rephrase, the letter of the law matches the characteristics of the resilience of this industry, with a conscious acceptance that the very core of the business, and the inherent associated risk, cannot change.

Another example is provided by the electricity power supply in the Northern US. This power supply system is known to be severely at risk after multiple failures occurred in the past three winters: however, the plan for correctives action taken by the US government is a prudent compromise that preserves the present balance and resilience, taking most initiatives at the local regional level and not at the national, or even at the continental level with Canada.

Ultra-Safe Systems

The fourth and last category of systems corresponds to those for which the risk of multiple fatalities is so high that even single accidents are unacceptable. If one accident occurs to one operator it may mean the end of business for all operators. This is the case of the energy industry (chemical, nuclear), and public transportation. The resilience of such systems adds to the characteristics seen in the previous category the need for a transfer of safety authority at the nation level or at international level, often leading to the creation of new agencies. Safety becomes a high priority at all levels of the system and an object of action by itself. Here the watchword of safety is *supervision*. Supervision means full traceability by means of information technology (IT) to enforce standardisation and personal accountability for errors, and growing automation to change progressively the role of technical people, freeing them from repetitive, time consuming and error prone techniques, and making them more focused on decisions.

For instance, advances in IT and computers provided aviation with an end-to-end supervisory system using systematic analysis of on-board black boxes, and leading to the eventual deployment of a global-sky-centralised-automated-management-system called the data-link. The impact on people and their jobs, whether pilots, controllers, or mechanics, is far-reaching.

To conclude this section, there are multiple examples in the industry which show that safety managers and professional are perfectly aware of how the jobs can be done more safely, yet fail to do it in practice. In most cases, they are limiting their safety actions because the system can survive economically with remaining accidents, and because the constraints associated with a better safety would not be affordable. This is a typical logic of losers and winners. As long as the losers stay isolated in a competitive market and their failures do not contaminate the winners, the system is not ready to change and give priority for resilience to competitiveness.

When the reality of accidents not only affects the unlucky losers but also propagates to the profession or to politicians, the system tunes resilience differently and increases the priority on safety. The first move consists in keeping the control on performance but increasing transparency on risk management. The second move is transferring the ultimate authority of risk management far from the workplace, giving gradual power to national and international agencies. Only very robust, organised, and money making systems may resist this last move, which generates an escalation of safety standards with little consideration for their cost.

Figure 16.3 illustrates a revision of Reason's general framework model of safety and resilience to fit with this idea of a cascade of resiliencies, with a paradoxical inversion of the paradigm. The spontaneous initial stage of resilience corresponds to highly competitive and relatively unsafe systems. When this stage is no longer viable, the system reverts to the next type of resilience, accepting more constraints and regulation, better safety, but losing some capacity to exceptional performance, etc. The cascade reversions continue until the system has reached a plateau where no further solutions exist to improve safety. This is generally the time of the death of the system, followed by the emergence of a new system (see section below). Note that the longer a system can remain in the first stages of resilience with high performance and low safety values, the longer its total life cycle will be (in that case, the system extends its life cycle, and postpones the arrival of the final asymptote).

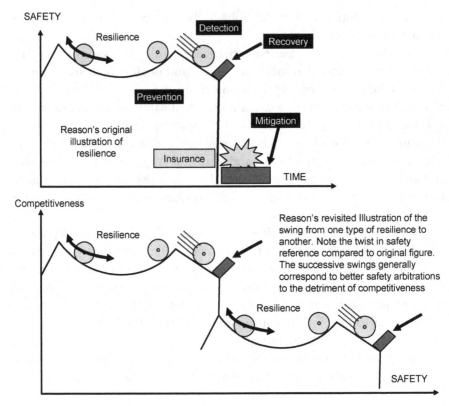

Figure 16.3: Revisiting Jim Reason's illustration of system resilience, turning to a cascade model of resiliencies

Understanding the Transition from One Type of Resilience to Another

As noted above there are two families of causes, external and internal, that may provoke the transition of one stage of resilience to another. The reader must understand the paradox of this concept of shift. Resilience, indeed, is by itself a concept that expresses the reluctance or resistance to shift. Therefore, each of the stages or types of resilience described in the previous section, oppose the shift as long as possible. However, no stage of resilience can exist forever in a given system. The external and internal forces will sooner or later provoke an inadequacy of this resilience system and ask for a new one. The conditions for these breakdown periods are described below.

External Causes of Transition

The external destabilising factors are in that case sudden and unpredictable: they can be a specific accident, or an economical crisis.

For example, a series of *listeriosis* infections occurred in France in the early 1990s due to improper preparation and storage of farm cheeses and milk products by farmers or (very) small food companies. Six young pregnant women died in 1992 and 1993, and the reaction of the media was so strong that this accident became the start of a large change in the farm product business. Within a period of two years after the problem occurred, a new law prescribed restrictions for farmers' food production, conservation, and conditions of food presentation in the marketplaces. Within a period of five years after the problem, a new national agency, the AFSSA ('Agence Française de Sécurité Sanitaire des Aliments'), was created (April 1999) concentrating the authority for food quality and control, and reinforcing the requirements for inspections (new directive dated May, 16th, 2000). From that date, the business drastically changed. Most isolated small fresh food companies collapsed or had to merge or sell activities to bigger trusts. The traditional French farm business that had bet for years on a great and cheap variety of products has now cut down the variety of products by 30%, been re-concentrated in the hands of a big combination of collective unions, and has definitively changed its working methods, moving to a new setting of resilience giving greater priority to safety.

However, the numerous examples given in the first section of this paper (e.g., Concorde, AZF follow-up actions) tend to demonstrate that such a reactive shift after an accident is far from being a standard in industry. Precisely because each type of human activity has adopted a system of resilience, it can long resist perturbations and accidents. In general, the very forces that push the professional activities to adopt a different system of resilience are more likely to be part of the history of the profession. This is why it is important to consider a more global framework model on the evolution of the life cycles of industry.

Internal Causes of Transition

The Life Cycle of Industry and Safety Related Concepts. Socio-technical systems, as well as humans, have a limited life span. This is specifically true if we adopt a teleological point of view where a social system is defined as a solution (a means) to satisfy a function (an end) (see Rasmussen, 1997; Rasmussen & Svedung, 2000, Amalberti, 2001). The solution is usually based on a strategy for coupling humans to technology; one can use the term 'master coupling paradigm'.

For example, the master coupling paradigm for train driving is based on a train driver being informed of danger by means of instructions and signals along the railway. However, some suburban trains have already adopted a different master coupling, becoming fully automated, hence with no drivers on board.

The master coupling paradigm for photographs has been stable for one century based on argentic printing. Nevertheless, the system totally collapsed in a short period of time in the early 2000s, being replaced by the new business of digital cameras and photos. Note also that the function of taking pictures was integrally maintained for end users through this drastic change.

Another example of a change in the master coupling paradigm has been the move from balloons to aeroplanes in air transportation. For nearly half of century between 1875 and 1925, commercial public air transportation was successfully made by airship. The airship era totally collapsed in a short period after the Hindenburg accident in New York in 1936, and was immediately replaced by the emerging aeroplane industry. Again, this change full preserved the main function of the system (transporting passengers by air).

Note that the same aeroplanes and associated master coupling paradigm that replaced balloons are themselves at the end of their life cycle. The present cockpit design is based on the presence of front-line actors (pilots, controllers) who have a large autonomy of decision-making. The deployment of the aircraft traffic satellite guidance via data link system that is just around the corner will drastically change the coupling paradigm and redefine all of the front-line professions on board and across the board.

The total lifespan for a given working activity based on a master coupling technological paradigm is likely to be equal to a human lifespan: half a century in the 19th century, and now about a century. However, during the period of the life cycle, where the master coupling paradigm remains macroscopically stable, the quality of the paradigm is continuously improving. Each period corresponds to a series of safety characteristics and resilience. This vision is another way to look at the different step of resiliencies that have been described above. We have seen that there are several categories of resilience with associated safety properties. We have seen that they are some conditions that lead to a transition from one to another. In a certain sense, we rephrase the idea from that section only adding the unavoidability of these transitions within the life cycle of systems. All system will transition and finally die. Some external events may precipitate this cycle that will inevitably come. The most paradoxical result is that speeding up the cycle will result in rapidly improving safety, but rapidly exposing the system to dying.

It is also important to consider that the usual way to look at system resilience and safety proceeds much more by audit and instant 'snapshots'.

With the information contained in the previous section of the chapter, such audits make it possible to classify the observed system in one of the four categories of resilience. It is then possible to infer the pragmatic behaviour after accidents, the limit of requirements that can be reasonably put on this system, and certain conditions that clearly announce the looming need to transition to the next category of resilience.

With the information presented in the following section of the paper, it will be possible to go beyond the instant classification and create a perspective of the dynamic evolution and the long time evolution of the system resiliencies. Four successive periods seems to characterise the life cycle of most systems (see Figure 16.4).

A short, initial period corresponds to the pioneering efforts. This is typical of the pre-industrial time where the master coupling paradigm is elaborated. At that stage, the specimens of the system are very few and confined to laboratories. Accident are very frequent (relative to the low number of existing systems), but safety is not a central concern, and pioneers are likely to escape justice when making errors.

A long period of optimisation can be termed the 'hope period'. Hughes (1983, 1994) says that pioneers are fired at that stage and replaced by businessmen and production engineers. The system enters into a long continuous momentum of improvement of the initial master coupling paradigm to fit the growing commercial market demand. Safety is systematically improved in parallel with the technical changes. It is time for hope: 'The upgraded version of the system already planned and available for tomorrow will certainly avoid the accident of today'. Despite remaining accidents, very few victims sue the professionals. As long as clients share the perception that the progresses are rapid, they tend to excuse loopholes and accept that errors and approximations are definitively part of this momentum of hope and innovation. The safer a system is, the more likely it is that society will seek to blame someone or seek legal recourse when injuries occur. For example, it is only in recent times that there has been an acceleration of patients suing their doctors. In France, for example, the rate of litigation per 100 physicians has increased from 2.5% in 1988 to 4.5% in 2001 (source MACSF).

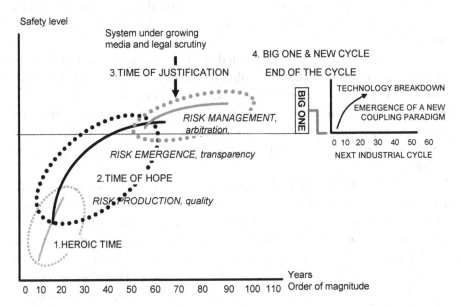

Figure 16.4: Evolution of risk acceptance along the life cycle of systems

The third period of the life cycle corresponds to the 'safety period'. The perception of the public is now that the system is reaching an asymptotic level of knowledge corresponding to a high level of potential performance. The customers expect to reap the benefits of this performance and do not hesitate any longer to sue workers. They particularly sue failing workers and institutions any time they consider that these people have failed to provide the expected service owing to an incorrect arbitration of priority in applying the available knowledge. A step further in this escalation of people suing workers is the shift from requirements for means to requirements for ends. People require good results.

The increase of legal pressure and media scrutiny is a paradoxical characteristic of systems at that stage. The safer they get, the more they are scrutinised. The first paradox is that the accidents that still occur tend to be more serious on account of the enhanced performance of the systems. Such accidents in safe systems are often massively more expensive in terms of compensation for the victims, to such an extent that in many sectors they can give rise to public insurance crises. Accidents therefore become intolerable on account of their consequences rather than their frequency. The skewing of policy caused by the interplay of technical progress and the increasing intolerance of residual risks is problematic. On the one hand the system experiences higher profits and better objective safety, but, on the other hand, infrequent, but serious, accidents cause an overreaction among the general public, which is capable of censoring and firing people, sweeping away policies, and sometimes even destroying industries. It is worth noticing that these residual safety problems can often bias objective risk analysis, assigning a low value to deaths that are sparsely distributed though they may be numerous, and assigning a high value to cases where massive concentration of fatalities occur even if they are scarce, such as aircraft accidents, problems of blood transfusion in France, or fire risks in hospitals. In essence, 100 isolated, singular deaths may have far less emotional impact than 10 deaths in a single event.

The consequence of this paradox is, for example, that patient injuries that occur tend to be massively more expensive in terms of patient compensation, and thus fuel the liability crisis.

The very last period of the life cycle is the death of the present coupling paradigm, and the re-birth of the system with a new coupling paradigm. The longer the life cycle of a system has lasted, the smaller will be the event causing its death. The last event that causes the system to die can be termed the 'big one'. The social life cycle of a system and of a paradigm can be extended for years when the conditions to shift and adopt a new system are not met. The two basic prerequisites for the change are the availability of a new technical paradigm and an acceptable cost of the change (balance between insurance and social crisis associated to the aging present system vs. cost of deployment of future system). Paradoxically, the adoption of the new coupling paradigm starts a new cycle and gives new tolerance to accidents, with a clear step back in the resilience space, associated with a relative degradation of safety.

Changing the Type of Resilience, Changing the Model of an Accident. When the safety of a system improves, the likelihood of new accidents decreases (see Table 16.2). The processing of in-service experience needs an increasingly complex recombination of available information to imagine the story of the next accident (Amalberti, 2001; Auroy et al., 2004).

The growing difficulties in foreseeing the next accident create room for inadequate resilience strategies as the system becomes safer and safer. To rephrase, the lack of visibility make resilience in safe systems much more difficult than in unsafe systems. This is also a factor that explains the usual short duration of the last stage of resilience in the system's life cycle. When systems have become ultra safe, the absence of ultimate visibility on a risk system may lead them towards their own death.

Table 16.2: Evolution of the prediction model based on past accidents

Up to 10^{-3}	10^{-3} to up to 10^{-5}	10^{-6} or better
The next accident will repeat the previous accidents	The next accident is a recombination of part of already existing accidents or incidents, in particular using the same precursors	The next accident has never been seen before. Its decomposition may invoke a series of already seen micro incidents, although most have been deemed inconsequential for safety

Conclusion: Adapt the Resilience – and Safety – to the Requirements and Age of Systems

The important messages of this chapter can be summarised by the following five points:

- Forcing a system to adopt the safety standards of the best performers is not only a naïve requirement but could easily result in accelerating the collapse of the system.

- Respect the ecology of resilience instead of systematically adopting forcing functions; the native or spontaneous resilience corresponds to well performing but unsafe systems, at least in the sense that safety is not a first priority.

- All systems will transition from their native resilience to new stages with much better associated safety. However, the ultimate stage will be so constraining that it will lead the system to die.

- It is crucial to have a good knowledge of the characteristics and the causal events announcing the transition for one resilience stage to another. These factors are external and internal.

- Smart safety solutions depend on the stage of resilience. Standardise, audit, and supervise are three key families of solutions that have to be successively deployed according to the safety level. Continuing standardisation beyond necessity will result in either no effect or in negative consequences.

Part III: Challenges for a Practice of Resilience Engineering

Chapter 17

Properties of Resilient Organizations: An Initial View

John Wreathall

Concept of Resilience

While formal definitions of resilience, and its associated field of application, resilience engineering, have yet to be developed, one of the simplest explanations is contained in the following description. Other equally valid definitions are discussed in other sections of this book. I do not claim primacy of this over others, but for this discussion, it suits my purpose.

> Resilience is the ability of an organization (system) to keep, or recover quickly to, a stable state, allowing it to continue operations during and after a major mishap or in the presence of continuous significant stresses.

The property in question of the organization is often safety, but should also include financial performance, and any other vital goal for the organization's well being.

From this description, we can see a significant difference from the more traditional techniques of safety management, like probabilistic safety assessment (PSA), accident root-cause investigations, and so on. First, PSA and similar methods are concerned with identifying and defending against a prescribed set of hazards using techniques that have significant limitations in terms of their ability to represent human and organizational influences appropriately – often the most important influences on safety performance. Second, many of the methods involve the analysis of events to identify 'causal' factors from root-cause analyses, fractions of events resulting from 'human errors' and so on.

While there are several techniques that allow these kinds of analyses, all suffer from various weaknesses. One is that the identification of causes of accidents tends to be a social as well as a technical process (in terms of what causes are considered 'acceptable' by the owners of the system). Another, that they are built around rather limited models of safety that ignore the roles of cultural and organizational influences, and, many times, rely on only partially recalled knowledge of events by participants. Most importantly, models describing the control of safety are built in isolation from other management activities, as if there was no connection, rather than them being integrally entwined.

The concepts of resilience and the anticipated tools for resilience engineering are intended to address these weaknesses head on. Thus, resilience engineering is a new management discipline that encompasses both safety management and other types of management, particularly process and financial management.

Approach of Resilience Engineering

If resilience is to ensure the organization keeps (or recovers to) a safe stable state, there are several processes that must go on to accomplish this goal. The purpose of resilience engineering is to develop and provide the tools for these processes. While the development of these is yet to be specified in detail, the following are the kinds of tools that would be developed.

Tools to Reveal Safety Performance

Organizations constantly struggle to understand how they are performing with regard to safety. 'Too much' safety is thought to limit the potential for operational profits, and too little will result in harm to the workers, lost production, and even loss of expensive facilities. While, in practice, safety and production are not necessarily in opposition – seeking to eliminate unwanted deviances in operation will improve both safety and production, for example – a lack of knowledge of the current levels of safety performance can lead organizations to take more conservative decisions than would be appropriate about production activities, or the organization's safety performance can 'drift' with a consequence that surprising accidents happen – as has been cited in the Challenger Shuttle accident.

Of course, organizations try to measure their safety performance, both in terms of industrial (worker) safety and process safety (accidents affecting the public and the environment). While many industries use different processes, some high-performing facilities (for example, in the nuclear industry) use essentially the same approach for both. Regardless of the processes used, both areas of safety require management processes based on performance data. Some of the most common processes involve trending safety outcomes like worker fatal accidents and lost-time injuries, costs of damage to process equipment, releases to the environment, and so on. Other industries, with which I am most familiar, such as nuclear power and chemical process industries, use safety modeling techniques such as probabilistic safety assessments to identify likely contributors to accidents (such as failures of protection equipment) and then trend the frequencies and durations for which such equipment is not operational. Some industries use techniques that involve some kinds of formal performance models and data-based evaluations of safety performance, including the aviation and defense sectors (although PSA is used in both of these, too). However, these are all very limited as sources of *operational management information*.

First, historic data from accidents will always be out-of-date as a measure of *today's* performance since data are from relatively rare events and almost always aggregated over long periods of time. They represent the consequences of decisions usually made significantly (years) earlier. The interpretation of these types of data is often uncertain, often biased by inevitable pressures to find simple explanations that are 'politically acceptable' and based on overly simple 'models of safety' that seek to identify one or two 'root causes' of accidents, neglecting the complexity of workplace pressures – see discussions by Hollnagel (1998; 2004), Dekker (2002) and others.

Second, PSA and similar models are very static interpretations of how accidents occur, focusing usually on simple descriptions of how combinations of hardware faults combine to cause bad outcomes and neglect the complexity and interactions seen in complex accidents. Reviews of PSAs, when compared with major accidents, show that they typically fail to identify underlying organizational processes that override the usual assumptions of independence of failures, and neglect the complexities of human behaviors usually involved in accidents, often treating humans as if they were simple machines.

This is not to say the PSA is of no use – it is often an excellent tool for evaluating designs and selecting between alternatives. Additionally, serious efforts are underway to remove some of the poor modeling of human performance, such as the development of the ATHEANA method (NRC, 2000). However, as a source of information about making operational decisions *now*, these methods, even the improved ones, are critically limited.

What are required are data that allow the organization's management to know the current 'state of play' of the safety performance within the organization, without suffering the problems outlined above.

Work has started to identify different data types and sources to provide the needed management information. An example is the work to develop leading indicators of organizational performance (see, for example, Wreathall, 1998, 2001; Wreathall & Merritt, 2003) that has been undertaken in several industries, including US nuclear power, aviation and oil exploration. This approach looks for data both at the working level (such as factors causing safety problems now for workers) and in organizational behaviors that can set them up to have vulnerabilities. These methods contribute to ensure the convergence of resilience at different levels of the organization in the management of uncertainties.

There are two tools that are used in a complementary manner in other industries that are used to measure the leading indicators associated with performance at the 'sharp end.' The first tool is intended to measure currently the kinds of problems commonly found in event investigations and near-miss reports. In most applications we have found 8–12 workplace and task factors typically encapsulate the dominant contributions to performance problems (Reason et al., 1998); these are identified by reviews of event reports and interviews with front-line workers. Typical examples seen across other industries include interfaces with other groups, lack of (or deficiencies in) relevant and timely input information, shortages of tools or other specific resources, and inadequate staffing. These problems are assessed proactively so that the organization does not have to wait for, and then suffer, the various costs of even partial failures in mission performance.

Rather, the accumulation of data allows management to take countermeasures before the problems cause failures. This tool simply solicits data via a web server from samples of workers on a periodic basis about how much each of these factors has been a problem in getting work done in a recent period of time.

The second tool is based on models of organizational effectiveness that focus on the core processes by which an organization accomplishes its mission. This will be based on the approach we have developed for other industries, the nuclear industry initially and more recently the rail and medical industries. This approach is based on work by Reason, who performed a review of about 65 different models that describe in various ways the relationship of organizational processes for successful and safe outcomes. The results of this review were to identify a set of common themes that collectively encompass the kinds of processes that are critical to organizational success in both safety and production through proactive risk management, as described by Wreathall & Merritt (2003).

The themes identified in the review are management commitment, awareness, preparedness, flexibility, reporting culture, learning culture, and opacity. Each of these themes has a particular meaning or significance in a different application domain. By customizing each of these themes for a particular domain, we can identify the potential sources of data from which managers at different levels (but aimed particularly at senior management) can assess the levels of performance and riskiness within their organization. It is important to note that, in most applications, very few (if any) new sources of data are needed; rather, it is a question of selecting the existing data that are particularly cogent for each of the themes and their customized form in a single domain.

The seven themes in highly resilient organizations are:

• *Top-level commitment*: Top management recognizes the human performance concerns and tries to address them, infusing the organization with a sense of significance of human performance, providing continuous and extensive follow-through to actions related to human performance, and is seen to value human performance, both in word and deed.

• *Just culture*: Supports the reporting of issues up through the organization, yet not tolerating culpable behaviours. Without a just culture, the willingness of the workers to report problems will be much diminished, thereby limiting the ability of the organization to learn about weaknesses in its current defences.

- *Learning culture*: A shorthand version of this theme is 'How much does the organization respond to events with denial versus repair or true reform?'
- *Awareness*: Data gathering that provides management with insights about what is going on regarding the quality of human performance at the plant, the extent to which it is a problem, and the current state of the defences.
- *Preparedness*: 'Being ahead' of the problems in human performance. The organization actively anticipates problems and prepares for them.
- *Flexibility*: It is the ability of the organization to adapt to new or complex problems in a way that maximizes its ability to solve the problem without disrupting overall functionality. It requires that people at the working level (particularly first-level supervisors) are able to make important decisions without having to wait unnecessarily for management instructions.
- *Opacity*: The organization is aware of the boundaries and knows how close it is to 'the edge' in terms of degraded defenses and barriers.

An example is provided at the end of this chapter of how a product of work performed to develop leading indicators of organizational performance can be adapted to reflect the kinds of issues of interest in the development of resilience engineering.

Other techniques have been developed to measure safety culture within organizations and their impact on safety. See, for example, work by Flin et al., relating to such work in the oil industry (Mearns et al., 1998; Mearns et al., 2003). The need is now to tie this approach to the concepts of resilience, to provide knowledge inside the organization about what its levels of safety are now.

Resources and Defenses

As well as knowing what is the present state of safety in the organization, it is important that the organization has available appropriate levels of resources (particularly reserves) that can react to sudden increasing challenges or the sudden onset of a major hazard – Reason has referred to this capability as providing 'harm absorbers' – analogous to shock absorbers in mechanical systems. These resources can be material, such as providing additional staff to cope with significant challenges (e.g., dedicated emergency response teams), or they can be design-oriented, such as building in additional times for people to react (some have called this 'white time') so that plant and management personnel have time to reflect on the nature of the challenge and take appropriate responses.

While appealing in the abstract, these concepts need development to answer appropriate management questions, like 'What kinds of resources to I really need?', 'How much in the way of resources, and at what cost?', 'When do I decide to deploy these resources?', 'What is the trigger?', 'How do I know my design 'white time' is adequate, and what am I giving up?'

Accompanying these questions is the overriding operational question: 'When and how do I decide I should sacrifice productivity for safety?' For example, it is when the production pressures are increasing that the need for greater safety questioning becomes more important. How can this be accomplished appropriately? How can the knowledge gained from the measures discussed above become useful?

On the matter of defenses, of course one class of defenses exists in the form of all the barriers that are built in to the design, and are represented in PSA models and the like. These have been extended by people such as Hollnagel (2004) to include more abstract (non-visible) barriers, such as standards, codes of conduct and procedures, and the like. However, equally important in the context of resilience is the role of people to act as positive promoters of safety. An example that has been identified in the world of medicine is where practices have evolved over time to cope with crises through changes in performance that are subtle (almost non-observable to the untrained eye). For instance, in observing a high-risk operation where the patient was suffering massive bleeding (a liver transplant), the anesthesiology team transformed from a routine of monitoring blood pressure and maintaining a regular blood transfusion supply, to a crisis response team that added staff to the team (*resources* discussed above), coordinated their efforts to add substantial amounts of blood in a short time (without jeopardizing safety by [e.g.,] skipping blood type checks) with virtually no orders or commands, and, when the patient was stable, relaxing back to the earlier behavior. This was all accomplished in a quiet, low-key manner.

From a safety perspective, the behavior avoided what in PSA terms would have been an initiating event. This positive performance was the result of the constant challenges faced in surgery (this procedure is notorious for the amount of blood loss and therefore people are not surprised by the need for response), and involved well practised performance in the face of a substantial challenge (with roles and duties well understood). However, the positive side of human performance, of which this is one example, is rarely or never factored into formal safety analyses and little forethought is given in preparing for such performance. How can the organization prepare and take credit for such performance? What infrastructure is needed? What needs to take place for this behavior to be 'normal'? Work is needed to provide formalisms to understand how to use this kind of behavior in safety analyses. Resilience provides a framework to explore this approach.

Understanding of Work as Performed, Not as Imagined

All of the above concepts are concerned with designing and monitoring the work in the organization. What seems to be a key factor in each of these examples of resilience engineering, is to have a realistic understanding of how work is actually performed, and then engineering all the tools and processes to exploit the beneficial features of that work (as with the case of the anesthesiologists in the transplant event), and to remove systems, processes and artifacts that get in the way of work being performed safely and effectively – the data gathering from the workers about 'things that get in the way of working safely' is one example. This same need applies at the organizational levels as well as the workers' level. How does the organization actually accomplish its work and how does that impact safety? Hence the need for measures of organizational behavior. This was already identified above as a needed area for work in resilience. How can this be accomplished?

Systems engineering techniques exist for describing formally the behavior of organizations and how *in reality* the organization manages to accomplish its goals. Many of these techniques stem from the work of the soft systems modelers in the late 1970s and 1980s, such as the work by Checkland (Checkland, 1981; Checkland & Scholes, 1990). Key elements of this approach involve systematic analysis of certain key facets of the organization's behavior, such as its how its commercial and regulatory environments affect its processes and standards, and associated decision-making. Leveson, Wreathall and others are looking at how this approach is being connected to resilience, and how it interacts with elements like the use of safety performance measures and cultural dimensions, for example.

At the levels of the individual workers, there is a need to consider the role of new technologies and how they may affect safety, such as creating new forms of hazards. Cook, Woods, Hollnagel, Wreathall and many others have written about examples of where new technologies have been introduced in the belief that they will eliminate known 'human errors' only to find that the potential for new error types has been overlooked, and that the new 'error' is possibly worse than the ones being eliminated.

Summary

Creating resilience engineering involves the development of several elements to create a set of tools that can, together, be used to enhance safety in the face of constant stresses and sudden threats. Work has already started on many of these tools, though not necessarily within the framework of resilience. This includes:

- the development of organizational and other performance indicators that provide current and leading information on safety performance;
- data analysis related to safety culture and climate, and an understanding of how they relate to performance;
- observations about how work is carried out in the real world, both at the worker level and for the organization as a whole;
- the timing and extent of resources that are necessary for 'harm absorption';
- how work processes and human behavior act to make safety *better* through individual and small team activities, as well as act as sources of failures;

- improved understanding of decision-making when it relates to sacrificing production goals to safety goals, how to accomplish it, and what resources are involved.

What is needed now is multidisciplinary efforts to integrate these activities to provide an integrated body of knowledge and tools for management to take advantage of these ideas.

Example: Adaptation of Leading Indicators of Organizational Performance to Resilience Engineering Processes

Concrete examples of some of the seven top-level issues used in the development of these indicators associated specifically with resilience are:

- *Flexibility*: The stiffness of the decision-making in the organization, and its failures to respond in a timely manner to an increasing need for revising its response to the pressures of production to allow increased protection, is typical of an organization that will have safety-related problems. Such a problem has been seen in several major incidents where the organization has maintained its fixation on production when the indications of a safety concern have been clear. Examples include the repeated violations at the Millstone nuclear power plant that led to the Nuclear Regulatory Commission issuing a fine of over $2 million in 1997 – see NY Times *"Owner of Connecticut Nuclear Plant Accepts a Record Fine"* September 28, 1997.[1]
- *Opacity* (and its corollary, *observability*): The extent to which information about safety concerns are kept closely held by a few individuals has been identified by analysts as a characteristic of organizations that are being set up for problems. For instance, Weick et al. (1999) refers to 'collective mindfulness' as a characteristic of highly reliable organizations: collective

---

[1]   Available at http://www.state.nv.us/nucwaste/news/nn10210.htm

mindfulness includes that fact that safety issues and concerns are widely distributed throughout the organization at all levels.

- *Just Culture* (also *openness*): The degree to which the reporting of safety concerns and problems is open and encouraged provides a significant source of resilience within the organization. The justice embedded in the *just culture* leads to the organization *not* penalizing the bearers of bad news – the opposite from what has been seen in the Millstone example above.

- *Management Commitment*: The commitment of the management to balance carefully the acute pressures of production with the chronic pressures of protection is a true measure of resilience. Their willingness to invest in safety and to allocate resources to safety improvement in a timely, proactive manner is a key factor in ensuring a resilient organization.

Acknowledgments

The author wishes to acknowledge the support provided by NASA Ames Research Center under grant NNA04CK45A to develop resilience engineering concepts for managing organizational risk, and for the support of many in the nuclear power industry for the evolution of ideas described herein.

Remedies

Yushi Fujita

Knowing the existence of mismatches between reality and formality is the first step for better remedy. Enforcing rules without understanding the mismatches is not an effective remedy. Appropriate monitoring mechanisms are a prerequisite for knowing the existence of mismatches. So are appropriate evaluation mechanisms for understanding mismatches. These mechanisms should maintain independence from and authority over administrative mechanisms.

Respecting humans at the front end (e.g., operators, maintenance persons) is also a useful step towards better remedy. They are the ones who best know the demanding reality, and how to cope with it. Overall safety has to be ensured by evaluating their behaviors, and corrective actions must be taken if their reactions are too risky.

Chapter 18

Auditing Resilience in Risk Control and Safety Management Systems

Andrew Hale
Frank Guldenmund
Louis Goossens

Introduction

Discussion of the need for resilience in organisations is only of academic value, if it is not possible to assess in advance of accidents and disasters whether an organisation has those qualities, or how to change the organisation in order to acquire or improve them, if it does not already have them. Proactive assessment of organisations relies on one form or other of management audit or organisational culture assessment. In this chapter we look critically at the ARAMIS risk assessment approach designed for evaluating high-hazard chemical plants, which contains a management audit (Duijm et al., 2004; Hale et al., 2005). We ask the question whether such a tool is potentially capable of measuring the characteristics that have been identified in this book as relevant to resilience. In an earlier paper (Hale, 2000) a similar question was asked in a more specific context about the IRMA audit (Bellamy et al., 1999), a predecessor of the tool that will be assessed here. That paper asked whether the audit tool could have picked up the management system and cultural shortcomings which ultimately led to the Tokai-mura criticality incident in Japan (Furuta et al., 2000), if used before the accident happened. That analysis concluded that there were at least eight indicators in the management system that the hazards of criticality had become poorly controlled, which such an audit, if carried out thoroughly, should have picked up. It is of course then a valid question whether these signals would have been taken seriously enough to take remedial action, which would have led to sufficient change to prevent the accident. Responding effectively to signals from audits is also a characteristic of a resilient organisation; but this in turn can be assessed by an audit, if it looks, as did the IRMA audit, at the record of action taken from previous audit recommendations.

The IRMA audit and its close relation, the ARAMIS risk assessment method, have been developed during studies in which the group in Delft has participated, in collaboration with a number of others.[1] These studies have tried to develop better methods of modelling and auditing safety management systems, particularly in major hazard companies falling under the Seveso Directives. Whilst the term 'resilience engineering' has never been used in these projects, there are many overlaps in the thinking developed in them with the concepts expressed in the chapters of this book and in the High Reliability Organisation and Normal Accident literatures, which have fed the discussions about resilience.

It seems valuable to ask the same question again here, but now in a more general context and not in relation to one particular accident sequence. We therefore present a critical assessment of our audit tool in relation to the generic weak signals of lack, or loss of resilience that are discussed in this book. We confine ourselves in this chapter to an audit tool as a potential measuring instrument for resilience and do not consider tools for measuring organisational culture. This reflects partly the field of competence of the principal author, partly the fact that other authors in this book are more competent to pose and discuss that question. It also reflects, however, a more fundamental proposition that, if resilience is not anchored in the structure of management

---

[1]   Relatively constant in these consortia have been the Dutch Ministry of Social Affairs and Employment, The Dutch Institute for Public Health, Environment and Nature, the British Health & Safety Executive, the Greek Institute Demokritos and Linda Bellamy, working for a number of different consultancies. Other institutes which have taken part include INERIS in France and Risø in Denmark. The consortia have worked under a succession of European projects, including PRIMA, I-Risk and ARAMIS.

systems, which audits measure, it will not have a long life. Organisational culture measuring instruments tend to tap into the aspect of group/shared attitudes and beliefs, which may rest largely in the sphere of the informal operation of the management system. Culture will, in this formulation, always be a powerful shaper of practice and risk control, but a lack of anchoring in structure may mean that culture can change unremarked, causing good performance to become lost. We believe that measuring instruments for resilience will need to be derived from both traditions. This chapter only assesses the contribution of one of them, the audit.

In the next section a brief summary of the ARAMIS risk assessment and audit tool is given, followed by a discussion of the way in which such a tool could provide some proactive assessment of the factors identified as providing resilience in the organisation.

Structure of the ARAMIS Audit Model

The model used in the ARAMIS project (Duijm et al., 2004) can be summarised as follows.

Technical Model of Accident Barriers: Hardware and Behaviour

The technical modelling of the plant identifies all critical equipment and accident scenarios for the plant concerned and analyses what barriers the company claims to use to control these scenarios. It is essential that this identification process is exhaustive for all potential accidents with major consequences, otherwise crucial scenarios may be missed, which may later turn out to be significant. All relevant accident scenarios need to be defined not only for the operational phase, but also for non-nominal operation, maintenance and emergency situations. In the ARAMIS project only some of these scenarios were worked out in detail, but the principles of the model could be applied to all of them. In the audits of chemical companies under the Seveso Directive, the major consequences covered are those which arise from loss of containment of toxic and flammable chemicals, and which result in the possibility of off-site deaths. If we were to lower the threshold of what we consider to be major consequences, to single on-site deaths, or even serious injuries, the task of identifying the relevant scenarios becomes rapidly larger, but the principles do not alter. A major project funded by the Dutch Ministry of Social Affairs & Employment is currently extending this approach to develop generic scenarios for all types of accident leading to serious injury in the Netherlands (Hale et al., 2004).

The scenarios are represented in the so-called bowtie model (Figure 3.2) as developments over time. Inherent hazards such as toxic and flammable chemicals, which are essential to the organisation and cannot be designed out, are kept under control by barriers, such as the containment walls of vessels, pressure relief valves, or the procedure operators use to mix chemicals to avoid a potential runaway reaction.

If these barriers are not present, or kept in a good operating state, the hazard will not be effectively controlled and the scenario will move towards the centre event, the loss of control. This is usually defined for chemicals as a loss of containment of the chemical. To the right of this centre event there are still a number of possible barriers, such as bunds around tanks to contain liquid spills, sprinkler systems to prevent or control fire, and evacuation procedures to get people out of the danger area. These mitigate the consequences. In this modelling technique barriers consist of a combination of hardware, software and behavioural/procedural elements, which must be designed, installed, adjusted, used, maintained and monitored, if they are hardware, or defined, trained, tested, used and monitored if they are behavioural/procedural. These management tasks to keep the barriers functioning effectively throughout their life cycle are represented in the three blocks at the base of Figure 18.1.

It is already common practice in chemical design to assign a SIL (Safety Integrity Level) value for safety hardware. These SIL values give nominal maximum effectiveness values for the barriers, which can only be achieved if the barriers are well managed and kept up to the level of effectiveness which is assumed to be the basis of their risk reduction potential. In the ARAMIS project equivalent SIL values were assigned to behavioural barriers in order to use the same approach. The audit is designed to assess whether the specific barriers chosen by the company are indeed well managed. The modelling of the scenarios and barriers themselves also gives insight into how good the choice of barriers is which the company has made, and whether another choice could provide more intrinsic safety, e.g. to eliminate the hazard entirely by using non-toxic substitute chemicals. In other words it assesses the inherent safety of the whole system.

Modelling the Management of the Barriers

The management model on which the audit is based is structured around the life cycle of these barriers or barrier elements (Figure 18.1).

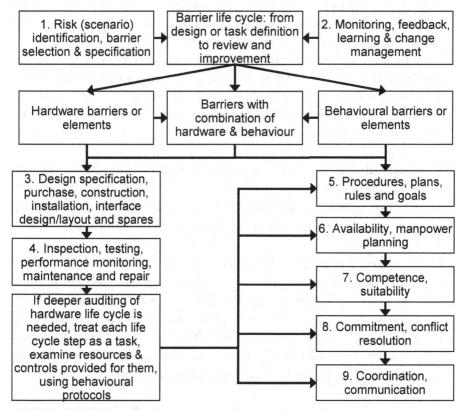

Figure 18.1: ARAMIS outline model of the barrier selection and management system

The life cycle begins in block 1 with the risk analysis process, which leads to the choice of what the company believes to be the most effective set of barriers for all its significant scenarios. Auditing the management process here assesses how well the company's own choice of barriers is argued, and so provides a check on the technical modellers' assessment of the barrier choice. In order that the life cycle should be optimally managed, the company needs to carry out a range of tasks, which will differ slightly depending on whether the barriers consist of hardware and software, of behaviour, or of a combination of both. These are detailed below the central block at the top of Figure 18.1 and we return to them below. Box 2 represents the vital process of monitoring the state and effectiveness of the barriers and taking corrective, or improvement action, if that performance declines, or new information about hazards or the state of the art of control measures, is found from inside or outside the company.

The life cycle tasks for hardware and software are shown to the left of Figure 18.1 (boxes 3 and 4) and these for behaviour and procedures are shown on the right in boxes 5 to 9. These two sets of tasks are, however, in essence fulfilling the same system functions. The barrier (technical or behavioural) must be

- specified (design specification or procedural or goal-based task description);
- it must be installed and adjusted (construction, layout, installation and adjustment, or selection, training, task allocation and task instruction);
- it must operate, a relatively automatic process for pure hardware/software barriers, but a complex one for behavioural barriers, since motivation/commitment and communication and coordination play a role, alongside competence, in determining whether the people concerned will show the desired behaviour; and
- it must be maintained in an effective state (inspection & maintenance, or supervision, monitoring, retraining).

Each step in ensuring the effectiveness of the hardware barriers (installation, maintenance, etc.) is itself a task carried out by somebody. If we want to go deeper into the effectiveness of this task we can ask whether the task has been defined properly (a procedure or set of goals), the person assigned to it has been selected and trained for it, is motivated to do it and communicates with all other relevant people, etc. In fact we are then applying the boxes 5 to 9 at a deeper level of iteration to these hardware life cycle tasks. Figure 18.1 shows this occurring for a deeper audit of the tasks represented by the boxes 3 and 4, the two halves of the life cycle for hardware/software barriers and barrier elements. If this iteration is carried out, boxes 5 – 9 are applied this time to tasks such as managing the spares ordering, storage and issuing for maintenance or modifications, or managing the competence and manpower planning for carrying out the planned inspection and maintenance on safety-critical hardware barriers.

In theory we could also take the same step for the tasks to select and train people or motivate them to do the tasks. We end up then with a set of iterations that can go on almost for ever – a set of Russian dolls, each identical inside each other. This would make auditing impossibly cumbersome. Again we could ask the question how far such iterations should go in a safe (or resilient) company, and at what point the company loses its way in the iteration loops and becomes confused or rigid. We return to some aspects of this in the discussion of the risk picture below.

Dissecting Management Tasks: Control Loops

We call the sets of tasks described in the preceding section as 'delivery systems', because they deliver to the primary barriers the necessary hardware, software and behaviour to keep the system safe. Each delivery system can be seen as having a series of steps, which can be summarised as closed loop processes such as in Figure 18.2 (the delivery process for procedures and rules), which deliver the required resources for the barrier life cycle functions and the criteria against which their success can be assessed and ensure that the use of the resources is monitored and correction and improvement is made as necessary. Such protocols were developed for each of the nine numbered boxes in Figure 18.2 (Guldenmund & Hale, 2004). Table 18.1 gives more detail of the topics covered by the protocols for the nine boxes. The monitoring and improvement loops described in these protocols must be linked to those represented in box 2 of Figure 18.1, but we do not show the lines and arrows there, in order to avoid confusing the picture.

The processes and the protocols of steps within each process can be formulated generically, as in Figure 18.1. However, in the audit they are applied specifically to a representative sample of the actual barriers identified in the technical model. Hence questions are not asked about maintenance in general, but about the specific maintenance schedule for a defined piece of hardware, such as a pressure indicator, identified specifically as a barrier element in a particular scenario. The audit then verifies performance against that specific schedule. Similarly, training and competence of specific operators in carrying out specified procedures related to, for example, emergency plant shutdown is audited, not just safety training and quality of procedures in general. When the audit approach is used by the company itself, the idea is that, over time, all scenarios and barriers would be audited on a rotating basis, to spread the assessment load.

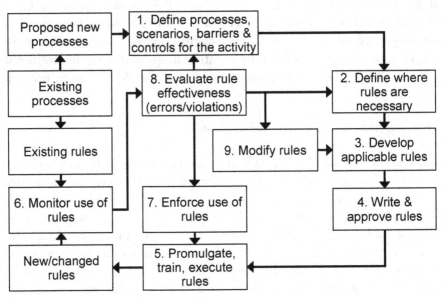

Figure 18.2: Example of management delivery process: procedures and rules

The audit tool provides checklist protocols based on the models outlined above. It has not been developed to the level of giving detailed guidance or extensive questions and checkpoints for the auditors. The assumption is made that it will be used by experienced auditors familiar with the safety management literature and able to locate the lessons and issues from that within the framework offered.

Further details of how the audit was used to rate management performance and to modify the SIL value of the chosen barriers can be found elsewhere (see Duijm et al., 2004. A description of all of the steps in the ARAMIS project will be published in 2006 in a special issue of the *Journal of Hazardous Materials*).

Table 18.1: Coverage of audit protocols

Protocol	Contents
Risk identification & barrier selection	Modelling of processes within system boundary, development of scenarios, (semi) quantification & prioritisation, selection of type and form of barrier, identification of barrier management needs, assessment of barrier functioning & effectiveness
Monitoring, feedback, learning & change management	Performance monitoring & measurement, on-line correction, feedback & improvement locally and across the system, design & operation of the learning system inside & outside the company, technical & organisational change management
Procedures, rules, goals	Definition & translation of high level safety goals to detailed targets, design & production of rules relevant to safety, rule dissemination & training, monitoring, modification & enforcement
Availability, manpower planning	Defining manpower needs & manning levels, recruitment, retention, planning for normal & exceptional (emergency) situations, coverage for holidays, sickness, emergency call-out, project planning & manning (incl. major maintenance)
Competence, suitability	Task & job safety analysis, selection & training, competence testing & appraisal, job instruction & practice, simulators, refresher training, physical & health assessment & monitoring
Commitment, conflict resolution	Assessment of required attitudes & beliefs, selection, motivation, incentives, appraisal, social control of behaviour, safety culture, definition of safety goals & targets, identification of conflicts between safety & other goals, decision-making fora and rules for conflict flagging and resolution
Communication, coordination	On-line communication for group/collaborative tasks, shift changeovers, handover of plant & equipment (production – maintenance), planning meetings, project coordination
Design to installation	Barrier design specification (incl. boundaries to working), purchase specification, supplier management, in-house construction management, type testing, purchase, storage & issues (incl. spares), installation, sign-off, adjustment
Inspection to repair	Maintenance concept, performance monitoring, inspection, testing, preventive & breakdown maintenance & repair, spares issue & checking, equipment modification

Does the Model Encompass Resilience?

As in our chapter about the resilience of the railway system (Chapter 9), we have defined eight criteria of resilience, which would require, as a minimum, to be measured by a management audit. These criteria are given below. After each criterion we indicate where in the ARAMIS audit system we believe it could, or should be addressed.

Table 18.2: Eight resilience criteria

1. Defences erode under production pressure.	Commitment & conflict resolution
2. Past good performance is taken as a reason for future confidence (complacency) about risk control.	Performance monitoring & learning.
3. Fragmented problem-solving clouds the big picture – mindfulness is not based on a shared risk picture.	Risk analysis & barrier selection, whole ARAMIS model
4. There is a failure to revise risk assessments appropriately as new evidence accumulates.	Learning
5. Breakdown at boundaries impedes communication and coordination, which do not have sufficient richness and redundancy.	Communication & coordination
6. The organisation cannot respond flexibly to (rapidly) changing demands and is not able to cope with unexpected situations.	Risk analysis. Competence.
7. There is not a high enough 'devotion' to safety above or alongside other system goals.	Commitment & conflict resolution
8. Safety is not built as inherently as possible into the system and the way it operates by default.	Barrier selection

For the purposes of this discussion we group these points in a way that links better with the ARAMIS model and audit. We use the following headings (numbers in brackets indicate where each of the above criteria is discussed):

A. Clear picture of the risks and how they are controlled (3, 8, elements of 5 and 6)
B. Monitoring of performance, learning & change management (elements of 1, 2 and 4)
C. Continuing commitment to high safety performance (1, 7)
D. Communication & coordination (elements of 5)

Of these, A and B are the most important and extensively discussed.

We shall conclude that the audit can cover many of the eight criteria in Table 18.2, but is weakest in addressing 2, 4 and 6.

In the sub-sections that follow we discuss the degree to which that assessment is indeed found and how the case studies used to test and evaluate the audit illustrated assessment of those aspects (see also Hale et al., 2005). We also discuss some of the problems of using the model and audit. These often raise the question whether the approach implicit in this assessment model is one that is compatible with the notion of resilience. As will be clear to the readers who have ploughed there way through the rather theoretical exposé in the description of the ARAMIS audit model, the model we use is complex, with many iterations. Resilience, on the other hand, often conjures up images of flexibility, lightness and agility, not of ponderous complexity. Our only response is to say that the management of major hazards is inescapably a complex process. Our models imply that we can only be sure that this is being done well if we make that complexity explicit and transparent.

Clear Picture of Risks and How they are Controlled

The essence of the ARAMIS method is that it demands a very explicit and detailed model of the risks present and the way in which they are controlled. It also provides the tools to create this. We believe that this modelling process is an essential step in creating a clear picture of how the safety management system works, which can then be communicated to all those within the system. Only then can they see clearly what part they each play in making the whole system work effectively. Trials with the approach (e.g. Hale et al., 2005) have shown that the explicitation of the scenarios and barriers, if done by the personnel involved in their design, use and maintenance, helps create much clearer understanding of what is relevant and crucial to the control of which hazards. This was a comment also made by the management of the companies audited in the ARAMIS case studies. They said they learned more from undergoing this audit than they had from other more generic system audits based on ISO-standards.

We devote considerable attention to this issue here, because we believe it is the most fundamental requirement for safe operation and for resilience. Without a clear picture of what is to be controlled, the other elements of resilience have no basis on which to operate.

Understanding How Barriers Work and Fail: How Complete is the Model? What the ARAMIS technique contains is explicit attention to the step from the barriers themselves to the management of their effectiveness, through their life cycle and the resources and processes needed to keep the barriers of different types functioning optimally. This direct linking of barriers to specific parts of the management system is unique in risk assessment and system modelling. We believe it provides a fundamental increase in the clarity of the risk control picture.

Of course the model is only as useful as it is complete. We know that a risk model can never be complete in every detail. We can hope that logical analysis of the organisation's processes, coupled with experience with the technology, can provide an exhaustive set of generic ways in which harm can result from loss of control of the technology. When we look at the major disasters that are often used to illustrate the discussion of high reliability or resilience, we do not see very often that the scenarios which eventually led to the accident were unknown. Loss of tiles on Columbia, the by-passing of the O-rings on Challenger, the starting of pumps on Piper Alpha with a pressurised safety valve missing, the mix of chemicals and water resulting in the gas cloud from Bhopal, etc. were all known routes to disaster, which had barriers defined and put in place to control them. What was lacking was enough clarity about the functioning of the barriers and about the commitment to their management to ensure that their effectiveness could be, and was, monitored effectively. At the level of the detail of how control is lost there will always be gaps in our model of the risk control system, as some of these disasters show. Sometimes these gaps are in the technical understanding of failure, such as of steel embrittlement at Flixborough, but more often the gaps are in our understanding of how people and organisations can fail in the conduct of their tasks. These failures sit in the delivery systems feeding the hardware and behavioural barriers. Vaughan's analysis (Vaughan, 1996) of the normalisation of deviance fits into this category, whereby subtle processes occur of becoming convinced that signals initially thought of as warnings are in fact false alarms. Normalisation of deviance is a failure in the monitoring and learning system (box 2 in Figure 18.1), which had previously not been clearly identified. Another example is the analysis of the classic accident at Three Mile Island nuclear reactor, in which operators misdiagnosed the leak in the primary cooling circuit, due to a combination of misleading or unavailable instrument readings and an inappropriate set of expectations about what were likely and serious failures. Until this accident, the modelling of human error had paid insufficient attention to such errors of commission, whose management falls under the design of instrumentation (box 3 in Figure 18.2) and the competence and support of operators (boxes 5 and 7).

The only recourse in the case of gaps in the model at this detailed level is then to rely on efficient learning (see 'Monitoring of performance, learning and change management' below); as new detailed ways of losing control of hazards are discovered within the organisation or comparable ones elsewhere, they need to be added to the model, as a living entity, reflecting the actual state of the system it models. This builds in resilience.

Functional Thinking. The ARAMIS models and the auditing techniques developed have a strong emphasis on functional process analysis. Functions are generic features of systems. How those functions are fulfilled in one specific company may be very different from another. Barriers also are defined functionally in the models, with the choice of how the barrier is implemented in a specific case left to the individual company. The quality of that choice is part of the audit process. To make the audit technique complete it will be necessary to build up and make available better databases of experience, indicating which choices of barriers in which circumstances are most likely to be effective (Schupp et al., 2005). Thinking in terms of functions provides a clearer picture of the essence of the risk control system, which does not get bogged down in the details of implementation. However, the audit must descend into these details to check that the function is well implemented in a particular company. It checks whether the managers and operators involved can see clearly not only what they specifically have to do (inspect a valve, monitor an instrument, follow up a change request, etc.) but how that fits functionally into the whole risk control system.

Control Loops & Dynamic Analysis. The idea of the life cycle of the barrier and the management of its effectiveness through the delivery of resources subject to their own control loops, introduces a dynamic element into the model, because these are processes with feedback. Resilience is essentially linked to a dynamic view of risk control. In this respect the Aramis model and audit matches the approach developed in Leveson's accident model (Leveson, 2004) and in the control-loop model of system levels put forward by Rasmussen & Svedung (2000). The emphasis of the dynamic modelling in the ARAMIS model has, however, been at the management level, a system level above the human-technology interactions which are the ones most described by

Leveson. Dynamic feedback loops at the management level are of course not new. The Deming circle has been a concept advocated for several decades and incorporated into the ISO management standards and their national forerunners. However, these standards and their auditors can be criticised for the fact that they have usually not paid enough attention to the detailed content of the feedback loops and the way they articulate with the primary processes to control risk. They have been too easily satisfied with the existence of bureaucratic and paper proof that planning and feedback loops exist, without checking whether their existence results in actual dynamic risk control and positive adaptation to hazard-specific new knowledge, experience and learning opportunities.

The use of dynamic models for the management processes presents problems of interfacing with the static fault and event tree models used by the technical modellers to quantify risk. A challenge for the future is to make technical risk modelling more dynamic, so that it can accept input from the control loops envisaged in such tools as ARAMIS. These are needed to get over to risk assessors and managers that risk is constantly varying due to changes in the effectiveness of the management and learning processes. A static picture of risk levels, established once and for all by an initial safety study, is an incorrect one, which encourages a lack of flexibility in thinking about risk control and hence of resilience.

Conceiving of Barriers as Dynamic. In ARAMIS and the more recent Dutch project looking at risk control in 'common or garden' accidents in industry[2], the control loop is linked with the functioning, and hence the effectiveness, of the barriers introduced to control the development of scenarios represented in the bowties. The link with barriers appears, according to the feedback from the case studies, to be one which can easily be communicated to people in the companies who are actually operating the processes and keeping the scenarios under control.

The original concept of barriers comes from the classic work of Haddon (1966) and has been used primarily in the old static, deviation

---

[2]   This project is called Workplace Occupational Risk Model or WORM (Hale et al., 2004) and involves the core participants mentioned in note 1.

models, typically codified in such approaches as MORT (Johnson, 1980). Barriers were originally conceived of largely as physical things in the path of energy flows. However, the concept has now been extended to procedural, immaterial and symbolic elements (Hollnagel, 2004), in other words everything that keeps energy (or more recently information) flows and processes from deviating from their desired pathways. In the work of the ARAMIS project we have extended this definition and given it a control-loop aspect. Barriers are defined as hardware, software or behaviour, which keeps the critical process within its safe limits. In other words, barriers are seen more as the frontier posts guarding the safe operating envelope, than as preventing deviation from the 'straight and narrow'. They are also defined as control devices, which must have three functions in order to be called effective barriers:

- Detect the need to operate,
- Diagnose what that control action is, and
- Carry out the control action.

We therefore do not see a warning sign or instrument reading as a barrier in its own right, only as a barrier element, since it is only effective if somebody responds to it and carries out the correct control action. In this sense we differ from Hollnagel (2004) by calling his symbolic and immaterial barriers only barrier elements. We therefore place behavioural responses more centrally in the definition of barriers, which we believe contributes to clearer thinking about them and to more resilience of the barriers themselves. A preliminary classification of 11 barrier types then indicates how each of these barrier functions is operationalised by an element of the barrier, either in the form of hardware/software, or of human perception, diagnosis or action (Hale et al., 2005), see Table 18.3.

This classification allows a direct link to be made between the choice of barrier type, the hardware, software and behavioural elements essential to its effective operation and the management functions needed to keep it functioning effectively (see Figure 18.1). A different choice of the type of barrier to achieve a given function can then be seen in the light of the management effort it brings with it, and the degree to which it can be thought of as inherently safe (criterion 8 of the resilience criteria in Table 18.2). By making the relative roles of hardware, software and behaviour very explicit, the technique forces the system controllers to think clearly about the prerequisites for effective functioning and to realise that all barriers depend for that effectiveness on human actions, either in direct use, or indirect installation, inspection, maintenance, etc. The issue of the resilience of the different barriers is therefore brought centre stage in the decision process of barrier selection and in auditing. At this point a judgement can be made as to how inherently safe the process is with its given barriers, or how narrow and fragile the margins are of the safe operating envelope.

Table 18.3: Classification of different barrier types.

	Barrier	Examples from chemical industry	Detect	Diagnose/ Activate	Act
1	Permanent – passive. Built into process	Pipe/hose wall, anti-corrosion paint, tank support, floating tank lid, viewing port in vessel	None	None	Hardware
2	Permanent – passive. Add-on barrier	Bund, dyke, drainage sump, railing, fence, blast wall, lightning conductor,	None	None	Hardware
3	Temporary – passive Put in place (and removed) by person	Barriers round repair work, blind flange over open pipe, helmet/gloves/safety shoes/goggles, inhibitor in mixture	None	None (human must put them in place, or put them on)	Hardware
4	Permanent – active	Active corrosion protection, heating/cooling system, ventilation, explosion venting, inerting system, bursting disc	None	None (operator may need to activate them for certain processes)	Hardware
5	Activated – hardware on demand	Pressure relief valve, interlock with "hard" logic, sprinkler installation, p/t/level control	Hardware	Hardware	Hardware
6	Activated – automated	Programmable automated device, control system or shutdown system	Hardware	Software	Hardware

	Barrier	Examples from chemical industry	Detect	Diagnose/ Activate	Act
7	Activated – manual Human action triggered by active hardware detection	Manual shutdown or adjustment in response to instrument reading or alarm, evacuation, putting on breathing apparatus or calling fire brigade on alarm, action triggered by remote camera, drain valve, close/open (correct) valve	Hardware	Human (S/R/K)	Human/ remote control
8	Activated – warned Human action based on passive warning	Wearing protective equipment in danger area, refraining from smoking, keeping within white lines, opening labelled pipe, keeping out of prohibited areas	Hardware	Human (R)	Human
9	Activated – assisted Software presents diagnosis to the operator	Using an expert system to assist diagnosis, alarm suppression system	Hardware	Software – human (R/K)	Human/ remote control
10	Activated – procedural Observation of local conditions not using instruments	(Correctly) follow start up/shutdown/batch process procedure, adjust setting of hardware, warn others to act or evacuate, (un)couple tanker from storage, empty & purge line before opening, lay down water curtain	Human	Human (S/R)	Human/ remote control
11	Activated – emergency Ad-hoc observation of deviation + improvisation of response	Response to unexpected emergency, improvised jury-rig during maintenance, fight fire using own initiative	Human	Human (K)	Human/ remote control

The coding S, R, K against human actions means the following:
S = skill-based, routine, well-learned responses which people carry out without needing to reflect or decide. They are triggered by well-known situations or signals.
R = rule-based, known and practised actions which have to be chosen more or less consciously from a repertoire of skills that have been learned in advance
K = knowledge-based improvisation when faced with situations not met before, or ones where there is no prescribed or learned response available to the person concerned and they have to make up a response as they go along.

Central or Distributed Control. All of these modelling efforts are designed to answer the need of the controllers in any system to have a model of the system they are trying to control, which shows them the levers they need to operate to influence the risks. In making these attempts at defining a central (umbrella) system model, we are of course in danger of falling into the trap that we are implying or even encouraging the view that control should also be situated centrally. Such a conclusion is not inevitable from building such an overall model. It is also possible that the functions in the total models are, in practice, carried out by many local controllers working in response to local signals and environments, but within one overall set of objectives. If control is distributed or local, there is, however, a need to ensure that the different actors communicate, either directly in person, or through observation of, and response to the results of other's actions. A central overview is essential for monitoring and learning, but not necessarily so for exercising control.

We need to resist the tendency to believe that central control is essential, since many studies have shown that decentralised control can be better and safer. An example of this was a recent study by Hoekstra (2001) of the possible implementation of the 'free flight' approach to ATM, in which control in certain parts of the air space is delegated from a central air traffic control centre to individual aircraft with relatively simple built-in avoidance algorithms. He showed that these could cope easily with traffic densities two orders of magnitude higher than the current maximum densities (which are already worrying en-route controllers) and could resolve even extreme artificial situations, such as a single aircraft approaching a solid wall of aircraft at the same altitude (with vertical resolution algorithms disabled), or 'super-conflicts' involving 16 aircraft approaching the same point in airspace from mutually opposing directions. (See also Chapter 4 by Pariès.) We need to distinguish the need for resilient systems to have both central overview and clarity for the purpose of system auditing, and local control for the purpose of rapid response.

The explicit risk control model may also be a help in coping with the ability to respond to unexpected situations and rapid changes, where distributed response is an essential element. However, this will only be so if the inventory of scenarios strives for completeness and extends to emergency situations and non-nominal system states. We have discussed above the impossibility of being fully comprehensive in detailed scenarios and the need for learning loops to be built in. The ARAMIS assessment methods are currently weak in dealing with the area of non-nominal situations, since there is nothing that forces the analyst to include these in the total model.

Monitoring of Performance, Learning and Change Management

In all of the management processes audited as 'delivery systems' in the ARAMIS audit there is an explicit step included, which is about monitoring the effectiveness of the resource or control delivered by that system. These are collected together in the learning and change management system in the ARAMIS model (Figure 18.3).

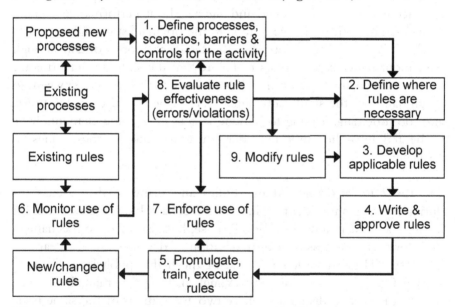

Figure 18.3: Management of learning and change

The ARAMIS audit emphasises the importance throughout the management system of having performance indicators attached to each step of all of the processes, both as indicators of state and performance (boxes 2 and 4) and of failure (box 3). Performance indicators are essential as the basis for deciding if defences are being eroded under pressure of production, routine violation or loss of commitment. The audit will indicate whether those performance indicators are in place, and also whether they are used, or whether a blind eye is being turned and performance is being allowed to slip.

Closely related to this are the systems for responding to changes in the external world (box 1: risk criteria, state of the art risk control knowledge, good practice) and for assessing proposed technical and organisational changes inside the organisation (boxes 6/7 and 8/9).

The audit also considers how all of the signals are used in the central boxes (5, 7, 9, 10), which assesses performance and proposed changes. At that point it is necessary that there is no complacency in interpreting good performance in the past as a guarantee of good performance in the future. The sensitivity of the learning system will also determine whether risk assessments are revised appropriately to take account of new evidence and signals about performance coming in. These are both resilience criteria in the original list of eight given in Table 18.2. It is critical here that the feedback loops contain enough data about relevant precursors to failure that the shifts in performance can be picked up. A crucial balance needs to be found between so much data that the significant signals cannot be sorted from the noise, and too little data, leading to blindness to trends. These elements have to be fed into the audit in order for it to pick up these clues to resilience.

The Two Faces of Change. Many studies and many of the discussions during the workshop identify that these feedback loops are essential for learning and adaptation in a 'good' sense, but also that the mechanism that drives the processes of normalisation of deviance are also learning processes. These have led to some of the highly publicised disasters, such as Challenger and Columbia (Vaughan, 1996, Gehman, 2003). The dilemma is how to distinguish these two applications of the same basic principles. The key to resilience is to be able to change flexibly to cope with unexpected or changed circumstances, but change itself can destroy tried and tested risk control systems.

The occurrence of violations of procedures is an example of a learning process gone wrong. Either the violations are unjustified and should have been corrected, or they are improvements in ways of coping, which should be officially accepted. In the specific protocol for assessing procedure management (Figure 18.2) there is a place for asking how this process of monitoring and changing rules is controlled. The only mechanism which we have come across, which explicitly addresses this issue, takes a leaf out of the book of technical change management and applies it to the management of procedural and organisational change. In one chemical company with an excellent safety record we found a mechanism in place whereby experienced specialists were available to modify standard procedures to cope with each application that deviated at all from standard, and in which the boundaries of that 'safe operation' of the procedures were very clearly defined. Within those boundaries operators and maintenance staff had freedom to adapt their controlling/operating behaviour. As soon as the boundary was reached the experts were called in to make the change, with an approval process involving more senior management at further defined boundaries of change. This process is reminiscent of the method of rule modification found by Bourrier in her study of a US nuclear plant (Bourrier, 1998). It operationalises a conclusion which we have reached from a number of our own studies that the crucial core of a rule management system is the system for rule monitoring and modification, not the steps of initial rule making. Such a view is proving to be a major culture shock when proposed in our railway studies, where the actors in the system are used to seeing rules as fixed truths to be carved in stone. Introducing an explicit rule change system is designed to make change more explicit and considered, so turning a smooth slide of rule erosion into a bumpy ride of explicit rule change.

Controlling Organisational Change. The chemical company mentioned above was also in the process of implementing a similar procedure for all organisational changes, such as manning changes, reorganisations, shifts of staff in career moves, etc. The idea of imposing an explicit control of organisational change also comes as a major culture shock in most organisations, even those operating in high hazard environments. They are used to seeing such reorganisations as unstoppable processes, which roll over them and cannot be questioned. The British Nuclear Installations Inspectorate has now imposed a rule requiring plants to present a detailed assessment of such organisational changes before they are approved (Williams, 2000). In order to implement such a rule, the company has to have a clear organisational picture of its risk control and management system. Then it can assess whether the reorganisation will leave any crucial tasks in that system unallocated. Reorganisation can also result in tasks being allocated to people not competent or empowered to fulfil them. It can also cut, distort or overload communication or information channels essential for local or overall control. The control of organisational change presents a challenge to industries and activities that have traditionally had an implicit risk control system and management (railways, ATM, etc.). Their picture of how their control system works may not be explicit enough to act as a basis for evaluating change (see also Chapter 3). This issue returns to the discussion raised in the section dealing with the 'clear picture of risks and how they are controlled'.

If we wish to evaluate organisational change in advance, this immediately raises the same questions as those which technical change management processes raise, namely when is a procedural or organisational change large enough to subject it to explicit scrutiny and demand a safety case to justify it? The chemical company mentioned above had begun to tackle this by making explicit such things as the (methods of calculating) minimum levels of manning[3] and competence for a range of activities, with an explicit approval process involving higher management if these were to be breached. This was a move appreciated by the operational and supervisory staff as a means to strengthen their hand in resisting subtle pressure from above to bend the rules informally. However, this is still largely unexplored territory for such companies and the lessons from accidents such as Challenger and Columbia show that the similar boundary posts and formal waiver procedures in place in those cases failed to stop the deviance normalisation processes there.

Continuing Commitment to High Safety Performance

The ARAMIS audit and its predecessor, the IRMA audit of I-Risk (Bellamy et al., 1999), have an explicit delivery system devoted to the delivery of commitment. This is combined in ARAMIS with the delivery of conflict resolution, a process separated out in the IRMA

---

[3] Earlier experience in posing this question to the association of chemical companies in the Rotterdam area of the Netherlands had led to a firm refusal to make such methods explicit and transparent or to develop common approaches to establishing them, with the argument that this was a matter for each company to decide in its own way and with professional secrecy, since it was linked to relative profitability.

audit. The difference between the two processes is as follows. Commitment applies particularly at the level of individuals and groups in the organisation and failures in it are concerned with violation of procedures, short cuts, neglect of routine checks, etc. This is the focus of the ARAMIS audit, which zooms in on on-line barrier performance. Conflict resolution deals much more with the conflicts emerging and controllable at the level of first-line, middle and senior management. These are concerned with such things as the decisions to curtail maintenance shutdowns for reasons of production, to reward managers on their quality or production figures despite poor safety figures, or to transfer resources for training from safety to security as a result of terrorist threats. They condition the choices at individual and group level, as does the individual commitment of the senior and top managers. Since these are the drivers behind the erosion of defences and the subordination of safety to other organisational goals, they are crucial protocols in assessing resilience. Unlike the management processes for managing the life cycle of hardware, or delivering manpower, competence or procedures, these delivery systems are much more diffuse in companies and therefore more difficult to audit. They approach most closely the traditional areas of safety culture, which can only be unravelled and understood by a combination of observation, interview and discussion. In the ARAMIS audit they are assessed by looking for explicit analysis of potential conflicts, by examining how safety is incorporated into formal and informal appraisal and reward systems, by looking for explicit instruments to monitor and discuss good and poor safety behaviour (violations), and by assessing senior management visibility and involvement in workplace safety. However, we accept that this is an area of the audit tool which is weak and which needs support from assessment methods drawn from research on culture and safety climate.

Communication and Coordination

What emerges from the studies of high reliability and resilience is a central emphasis on intensive communication and feedback, either for the individual or for and within the group which is steering the system within the boundaries of its safe envelope. It seems necessary to audit, or study in detail what it is that is taking place within this communication and feedback, in order to understand if it contains the requisite information and models to cope with the range of situations it will meet. What we are looking for is a better understanding of how this commonly described or demanded aspect of a culture of safety works in detail. What is observed? How is it interpreted? What is communicated, how and with whom? What are signals indicating a movement towards the boundaries of control? How are they distinguished from other signals? In the ARAMIS audit this aspect receives specific attention in the delivery system relating to communication and coordination. This concentrates on the communication within groups, particularly when their tasks are distributed in time or space. It therefore assesses the operation of barriers where combined action by control room and field staff is necessary, the transfer of control at shift changeovers and between operations and maintenance in isolating, making safe and opening plant for maintenance and reassembling, testing and returning it to service.

Conclusions and General Issues

In the previous sections we have indicated that a number of the major criteria for assessing or recognising resilience, which were listed in Table 18.2, are potentially auditable by a system such as the ARAMIS audit. The audit and its underlying modelling techniques pay a lot of attention to establishing a clear picture of the risks to be controlled and how they are eliminated, minimised or managed. The full range of possible scenarios, the well-chosen barriers, the well-managed barrier life cycles, the functioning feedback loops and the critically reviewed learning and adaptation processes are explicitly covered in the auditing system. So is attention to commitment, conflict resolution and communication.

Dependence on Auditor Quality

However, the assessment of all of these factors, as in all audit systems, depends very much on the quality and knowledge of the auditor and whether the relevant aspects are picked up and assessed. This depends in turn on the auditor's willingness and ability to judge the quality of arguments offered by the companies in justifying their approach to risk control and its management. Only highly experienced auditors, knowledgeable about the industry being audited can make such judgements. In the five case studies carried out to test the ARAMIS audit this issue was highlighted. The structure of the audit and its support is not mature enough yet to provide acceptable reliability in the assessments. The issue of the quality and experience of auditors presents problems for the regulators, whether these are government inspectorates or commercial certification bodies. Where can they attract such knowledgeable and valuable people from and how can they afford to pay them? The companies themselves, if they use such scenario-based auditing techniques, also need to commit themselves to putting high quality staff into these functions. These need to be people who can step back far enough from the familiar surroundings of the plant and its risk controls to assess them independently and critically. The central audit teams employed by a number of international chemical companies, who travel to company sites for audits lasting several days at regular intervals, seem to come close to fulfilling this essential role as critical testers of the resilience of the organisations they visit.

Marrying Auditing and Culture Assessment

The ARAMIS audit, but undoubtedly other audit tools, can provide the hooks on which to hang the questions that can attempt to assess the resilience of companies according to the criteria we have listed in Table 18.2. However, we have to admit that the questions to probe those issues are not always made explicit in the audits. If we wish to ensure that the questions do get asked in ways which will produce revealing answers, we will need to adapt the audit protocols to provide a much more explicit focus. In the ARAMIS project an attempt was made to start on this process, which would require a closer coupling between the traditionally separate areas of safety management structure and safety culture. Discussions to try to link specific dimensions of safety culture to specific aspects of safety management structure were carried out, but this failed to lead to any consensus. Further attempts are necessary to marry these two areas of work, which have arisen from separate traditions, one rooted in psychology and attitude measurement and the other in management consultancy, so that we can devise effective audit tools to assess both in an acceptable period of time.

What if Resilience is Measured as Poor?

An audit tool should, according to our arguments, be able to recognise at least a number of the weak signals indicating that an organisation is not resilient. The question then is how to improve matters. What the audit tells the organisation is which of the performance indicators is off-specification. If these performance indicators are widely spread throughout the risk control system, they should tell the company where the improvements must take place. Is that in the monitoring and review of performance? Is it in tackling significant amounts of violation of procedures? Is it in convincing staff that better performance is possible? Is it in the design and layout of user and maintenance friendly plant and equipment? Or is it in a move to risk-based maintenance? What an audit cannot say is how to achieve that change. The processes of organisational change, of achieving commitment and of modifying organisational culture require a very different set of expertise than that of the auditor. We leave the analysis of those change processes to other experts.

Chapter 19

How to Design a Safety Organization: Test Case for Resilience Engineering

David D. Woods

In the aftermath of the Columbia space shuttle accident (STS-107), the investigation board found evidence of an organizational accident as NASA failed to balance safety risks with intense production pressure (Gheman, 2003). Ironically, a previous investigation examining a series of failures in Mars exploration missions also focused on breakdowns in organizational decision-making in their recommendations (Stephenson et al., 2000). Both reports diagnosed a process where the pressure for production to be 'faster, better, cheaper', combined with poor feedback about eroding safety margins, led management inadvertently to accept riskier and riskier decisions.

Woods (2005a) links these accident analyses to patterns derived from previous results and argues that organizational accidents represent breakdowns in the processes that produce resilience. Balancing the competing demands for very high safety with real-time pressures for efficiency and production is very difficult. As pressure on acute efficiency and production goals intensifies, first, people working hard to cope with these pressures make decisions that consume or 'sacrifice' tasks related to chronic goals such as safety. As a result, safety margins begin to erode over time – buffering capacity decreases, system rigidity increases, the positioning of system performance relative to boundary conditions becomes more precarious (cf., Chapter 2). Second, when margins begin to erode as a natural response to production pressure, it is very difficult to see evidence of increasing or new risks. Processes that fragment information over organizational boundaries and that reduce cross-checks across diverse teams leave decision makers unable to recognize the big picture, that is, unable to reframe their situation assessment as evidence of a drift toward safety boundaries accumulates (until a failure occurs and with the benefit of hindsight the evidence of new dangers seems strong and unambiguous).

How do people detect that problems are emerging or changing when information is subtle, fragmented, incomplete or distributed across the different groups involved in production processes and in safety management? Many studies have shown how decision makers in evolving situations can get stuck in a single problem frame and miss or mis-interpret new information that should force re-evaluation and revision of the situation assessment (e.g., Johnson et al., 2001; Patterson et al. 2001). A recent synthesis of research on problem detection by professional decision makers (Klein et al., 2005) found that reframing is a critical but difficult skill. Reframing starts with noticing initial signs that call into question ongoing models, plans and routines. How do these discrepancies lead people to question the current frame? When do they become suspicious that the current interpretation of events is incomplete and perhaps incorrect?

The initial signs are always uncertain and open to other interpretations. These indicators can easily be missed or rationalized away rather than lead to questioning and revision of the current frame. For example, studies have shown that a skilled weather forecaster comes in to work searching for the problems of the day, which comprise the unsettled parts of the scene that will need to be closely monitored (Pliske et al., 2004). In other words, the expert adopts a highly suspicious stance to notice and pursue small discrepancies despite the workload pressures and attentional demands. Less-skilled forecasters are much more reactive given other demands and do not reserve time to pursue these small (usually unimportant) discrepancies. As this example indicates, factors related to expertise, workload, and attentional focus can all contribute to a tendency to become stuck in a single view or frame, even as evidence is accumulating that suggests alternate situation assessments (Klein et al., 2005).

A resilience perspective on accidents such as Columbia allows one to step away from linear causal analyses that become stuck on the proximal events in themselves, on red herrings such as human error, or vague 'root causes' such as communication. Major accidents, like Columbia, are late indicators of a system that became brittle over time, of a safety management process that could not see the increasing brittleness, and of safety management that was in no position to help line management respond to increasing brittleness. As a result, failures of safety management in the face of pressure to be 'faster, better, cheaper' reveal that more effective techniques should provide the ability:

- to revise and reframe the organization's assessment of the risks it faced and the effectiveness of its countermeasures against those risks as new evidence accumulates.
- to detect when safety margins are eroding over time (monitor operating points relative to boundaries as noted in Cook & Rasmussen, 2005), in particular, to monitor the organization's model of itself – the risk that the organization is choosing to operate nearer to safety boundaries than it realizes.
- to monitor risk continuously throughout the life-cycle of a system, so as to maintain a dynamic balance between safety and the often considerable pressures to meet production and efficiency goals.

The organizational reforms proposed by the Columbia Accident Investigation Board try to meet these criteria, which makes this accident report the first to recommend a resilience strategy as a fundamental mechanism to prevent future failures.

Dilemmas of Safety Organizations

Using a resilience approach to safety, I provided some input to the Columbia Accident Investigation Board (CAIB) which seemed consistent with the Board's own analysis and recommendation directions. Later Congress, as NASA's supervisor, wanted to check on NASA's plans to implement the CAIB's recommendations, especially the modifications to NASA's safety office. Congressional staffers asked several people to comment on the changes. As background I circulated a draft of my input to the board (what later evolved into Woods, 2005a). The staffers were very interested in this perspective, but to my surprise asked a simple and challenging question – how does one design a safety organization to meet these criteria? I was caught completely off guard, but immediately recognized the centrality of the question.

Resilience engineering, if it is a meaningful and practical advance in safety management, should be able to specify the design of safety organizations as a work-a-day part of the organization's activities.

The staffers' question put me on the spot. As always when confronted with a conceptual surprise my mind shifted to a diagnostic search mode: why is the job of a safety organization hard? The resilience paradigm suggested organizations needed a mechanism that questions the organization's own model of the risks it faces and the countermeasures deployed. Such a 'fresh' or outside perspective is necessary for reframing in cognitive systems in general. A review and reassessment was necessary to help the organization find places where it has underestimated the potential for trouble and revise its approach to create safety. A quasi-independent group is needed to do this – independent enough to question the normal organizational decision-making but involved enough to have a finger on the pulse of the organization (keeping statistics from afar is not enough to accomplish this).

Why is developing and maintaining this questioning role difficult and unstable? Because organizations are always under production pressure (though sometimes the pressure on these acute goals can be stronger or weaker), the dilemma for safety organizations is the problem of 'cold water and an empty gun.' Safety organizations, if they assess the organization's own models of how it is achieving safety, raise questions which stop progress on production goals – the 'cold water.' Yet when line organizations ask for help on how to address the factors that are eroding or reducing resilience, while still being realistic and responsive to the ever-present production constraints, the safety organization has little to contribute – the 'empty gun.' As a result, the safety organization fails to better balance the safety/production trade-off in the long run and tends to be shunted aside. In the short run and following a failure, the safety organization is emboldened to raise safety issues (sacrifice production goals), but as time flows on, the memory of the previous failure fades, production pressures dominate, and the drift processes operate unchecked (as has happened in NASA before Challenger and before Columbia, and can happen again).

From the point of view of managing resilience, a safety organization should monitor and dynamically re-balance the trade-off of production pressure and risk. The safety organization should see 'holes' in the organization's decision processes, reframe assessments of how risky the organization has been acting, to question the organization's assumptions about how it creates safety. How could a safety organization be designed to meet these ambitious goals since these are rather difficult cognitive functions to support in any distributed systems? Even worse, in order to avoid the trap of 'cold water and empty guns,' I was in effect asking the leadership of an organization to authorize and independently fund a separate group whose role was to question those leaders' decisions and priorities.

And then, if the safety organization was authorized and provided with an independent set of significant resources, it was committed to offer positive action plans sensitive to the limited resources and larger pressures imposed from outside. To accomplish this requires a means for safety management to escape the fundamental paradox of production/safety conflicts: safety investments are most important when least affordable. It is precisely at points of intensifying production pressure and higher organizational tempo that extra investments are required in sources of resilience to keep production/safety trade-offs from sliding out-of-balance. What does Resilience Engineering offer as guidance to better balance this trade-off?

The 4 'I's of Safety Organizations: Independent, Involved, Informed, and Informative

At this point I had used a resilience perspective to provide common ground for an exchange on the dilemmas of safety organizations. But I was still on the spot and the staffers were insistent, how can safety organizations be designed to cope with these dilemmas? How did successful organizations confront these dilemmas?

To help organizations balance safety/production trade-offs, a safety organization needs the resources and authority to achieve independence, to be involved, informed and informative. My response was that safety organizations are successful when they:

- provide an independent voice that challenges conventional assumptions about safety risks within senior management,
- have constructive involvement in targeted but everyday organizational decision-making (for example, ownership of

technical standards, waiver granting, readiness reviews, and
anomaly definition),

- actively generate information about how the organization is actually
  operating and the vectors of change that influence how it will
  operate (informed),
- use information about weaknesses in the organization and the gap
  between work as imagined and work as practised in the
  organization to reframe and direct interventions (informative).

These four 'I's provide a simple mnemonic that concisely captures
the difficulty in designing a safety organization: these four requirements
are in conflict! At best, the relationship between the safety organization
and senior/line management will be one of constructive tension. Safety
organizations must achieve independence enough to question the
normal organizational decision-making, provide a 'fresh' point of view,
and help the parent organization discover its own blind spots.
Challenging conventional assumptions of senior management limits the
voice as fresh views bring unwelcome information and seem to distract
from making definitive decisions or building support for current
management plans. Inevitably, there will be periods where senior
management tries to dominate the safety organization. The design of
the organizational dynamics needs to provide the safety organization
with the tools to resist these predictable episodes by providing funding
directly and independent from headquarters. Similarly, to achieve
independence, the safety leadership team needs to be chosen and
accountable outside of the normal chain of command.

Safety organizations must be involved in enough everyday
organizational activities to have a finger on the pulse of the
organization and to be seen as a constructive participant in the
organization's activities and decisions that affect the balance across
safety and production goals. In general, safety organizations are at great
risk of becoming information-limited as they can be shunted aside from
real organizational decisions, kept at a distance from the actual work
processes, and kept busy tabulating irrelevant counts when their
activities are seen as a threat by line or by upper management (for
example, the 'cold water' problem). Simply by being positioned to have
a voice at the top can leave the safety organization quite disconnected
from operations and exacerbate information limits. By being informed,
the safety organization can be informative, and the strongest test of this

criterion is the ability to identify targets for investments to enhance aspects of resilience and to prioritize across these targets of opportunity. To be constructive, a safety organization needs to control a significant set of resources and have the authority to decide how to invest these resources to help line organizations increase resilience and enhance safety while accommodating production goals. For example, the safety organization could decide to invest and develop new anomaly response training and rehearsal programs when it detects holes in organizational decision-making processes. Involvement, balanced with independence, allows the safety organization to provide technical expertise and enhance coordination across the normal chain of command. In other words, the involvement focuses on creating effective overlap across different organizational units (even though such overlap can be seen as inefficient when the organization is under severe cost pressure).

Balancing the four 'I's means that a safety organization is more than an arm's length tabulator, does more than compile a trail of paperwork showing the organization meets requirements of 'safety' as defined by regulators or accreditors, is more than a cheerleader for past safety records, and more than a cost center that occasionally slows down normal production processes. Being involved and informed requires connections to the character and difficulties of operations (the evolving nature of technical work as captured e.g., in the studies in Nemeth, Cook & Woods, 2004). Being independent and informative requires a voice that is relevant and heard at the senior management level. By achieving each pair and making them mutually reinforcing, safety management becomes a proactive part of the normal conduct of the organization.

The safety organization's mission then is to monitor the organization's resilience including the ability to make targeted investments to restore resilience and reduce brittleness. In reaching for the four 'I's, the safety organization functions as a critical monitor of the gap between work as imagined and work as practised and generates tactics to reduce that gap. As a result, the safety organization becomes a contributor to all of the organization's goals – by enhancing resilience both safety and production are balanced and advance together as new capabilities arise and as the organization faces new pressures.

Safety as Analogous to Polycentric Management of Common Pool Resources

The analysis above and the four 'I's as a potential solution to the challenge case parallels analyses of how complex systems avoid the tragedy of the commons (Ostrom, 1990; 1999). The tragedy of the commons concerns shared physical resources (among the most studied examples of common pools are fisheries management and water resources for irrigation). The tragedy of the commons is a name for a baseline adaptive dynamic whereby the actors, by acting rationally in the short term to generate a return in a competitive environment, deplete or destroy the common resource on which they depend in the long run. In the usual description of the dynamic, participants are trapped in an adaptive cycle that inexorably overuses the common resource; thus, from a larger systems view the local actions of groups are counterproductive and lead them to destroy their livelihood or way of life in the long run.

Organizational analyses of accidents like Columbia seem to put production/safety trade-offs in a parallel position to tragedies of the commons. Despite organizations' attempts to design operations for high safety and the large costs of failures in money and in lives, line managers under production pressures make decisions that gradually eat away at safety margins, undermining the larger common goal of safety. In other words, maybe safety can be thought of as an abstract common pool resource analogous to a fishery. Or, alternatively, dilemmas that arise in managing physical common pool resources are a specific example of a general type of goal conflict where different groups are differentially responsible and affected by different sub-goals, even though there is one or only a couple of commonly held over-arching goals (Woods et al., 1994, Chapter 4).

Developing the analogy further, the standard view of how to manage common pool resources is to create a higher level of organization responsible for the resource over its entire range and over longer periods of time. This organization then needs authority to compel individuals or local groups to modify their behavior sacrificing short term return and autonomy in order for the higher level organization to analyze and plan behaviors that sustain or grow the resource over the long term – a command organization. Safety management theory often seems to make similar assumptions and propose similar responses, i.e., a command structure is needed from regulators to companies or from management to line operations that takes a broader view and compels workers and line managers to modify behavior for a long term common good.

Ostrom (1999) reviews the empirical results on how people actually manage common pool resources and finds the standard view unsupported by the evidence. Basically, she found that overuse by local actors is not inevitable and that command style relationships across levels of organizations do not work well. Instead, she finds from research on co-adaptive systems that common pool resources can be effectively managed through polycentric governance systems. Polycentric systems provide for multiple levels of governance with overlapping authority in a dynamic balance but where there is no single governance center which directs or 'commands' unilaterally. Her synthesis of research identifies a variety of conditions and properties for polycentric management of common resources (such as cross-communication, shared norms, trust, and reciprocity; Ostrom, 2003).

The proposed four 'I's of safety organization design can then be seen as additional policy guidance for how to build effective polycentric management to balance multiple interacting goals. Achieving a dynamic balance across multiple centers of governance – some closer to the basic processes but with narrower field of view and scope of action and others farther removed but with larger fields of view and scopes of action, would seem to require a quasi-independent, intersecting organization that can cross connect these different levels of organization to be both informed and informative. By being outside a nominal chain of command, such groups can question and help revise assessments as evidence and situations change, as well as intervene with targeted investments to help resolve short term dilemmas (independent and involved).

Recent research on distributed cooperative systems made possible by new computer technology also seems to support the analogy, for example studies of the change to 'free flight' in managing the national air transport system support and extend Ostrom's findings (see Smith et al., 2004). The tools that have proved necessary to make collaboration work between air carriers and FAA authorities given new capabilities for communication at a distance and given the demands for adaptive behavior as congestion and weather change also provide other ideas for the design of polycentric management systems. Similarly, studies of how military organizations delegate authority to adapt plans to surprising situations provide lessons that also can be applied to guide polycentric management (e.g., Woods & Shattuck, 2000).

The analogy suggests that findings from managing physical common pool resources and findings from how goal conflicts between safety versus production are resolved (Woods et al., 1994, chapter 4) may converge and mutually reinforce or stimulate each other. For example common pool research may benefit from examining the reframing processes which are central to the resilience approach to safety under different management structures.

Summary

Organizations in the future will balance the goals of both high productivity and ultra-high safety given the uncertainty of changing risks and certainty of continued pressure for efficient and high performance. This organization will be able to (a) find places where the organization itself has missed or underestimated the potential for trouble and revise its approach to create safety, (b) recognize when the side effects of production pressure may be increasing safety risks and, (c) develop the means to make targeted investments at the very time when the organization is most squeezed on resources and time.

To carry out this dynamic balancing act, a new safety organization will emerge – designed and empowered to be independent, involved, informed, and informative. The safety organization will use the tools of Resilience Engineering to monitor for 'holes' in organizational decision-making and to detect when the organization is moving closer to failure boundaries than it is aware. Together, these processes will create foresight about the changing patterns of risk before failure and harm occur.

Acknowledgements

This work was supported in part by grant NNA04CK45A from NASA Ames Research Center to develop resilience engineering concepts for managing organizational risk. I particularly thank the congressional staffers who provided an opportunity to review NASA's post-Columbia reform plans and who challenged the concepts for achieving resilience. The ideas here benefited greatly from the inputs, reviews, and suggestions of my colleagues Geoff Mumford and Emily Patterson. The remaining gaps are my own.

Rules and Procedures

Yushi Fujita

Work rules and operating procedures are much less effective
than they are normally believed to be. Trying to fix the system
at an optimal point for extended time with work rules and
procedures may be a feeble idea. "Situated Cognition," a
school of sociology, argues that the idea of controlling work
with rules and procedures is only an administrative view. Often
times, critical decisions are (or need to be) governed by social
convenience and common sense. Work rules and procedures
that are tuned to prescribed scenarios may be too specific and
rigid, while those that can flexibly absorb fluctuating reality
may be too general and loose. What matters more is not the
way rules and procedures are specified, but the way in which
humans actually behave. Observation is a prerequisite for
knowing actual human behaviors and understanding the
background.

Chapter 20

Distancing Through Differencing: An Obstacle to Organizational Learning Following Accidents

Richard I. Cook
David D. Woods

The future seems implausible; the past seems incredible.
Woods & Cook (2002)

Introduction

A critical component of a high resilience in organizations is continuous learning from events, 'near miss' incidents, and accidents (Weick et al., 1999; Ringstad & Szameitat, 2000). As illustrated by the many cases referenced in this book, incidents and failures provide information about the resilience or brittleness of the system in the face of various disruptions. This chapter explores some of the barriers that can limit learning even by generally very high quality organizations.

Despite terrible consequences, accidents, as fundamentally surprising events, offer an opportunity for learning and change as there is a profound sense among many that the usual concepts and policies are insufficient to cope with what has happened (Woods et al., 1994). The immediate aftermath of a serious failure produces an atmosphere of inquiry and frees up resources normally dedicated to production, which are refocused on the accident and its consequences. The lines that divide participants, management, regulators, and victims from each other are momentarily thin. As one National Transportation Safety Board observer put it, "when the vividness of the tragedy is fresh in everyone's mind, and the broken wreckage is still smoldering ... people ... have only one pressing goal, and that is to determine precisely what happened and see that it does not happen again" (NTSB, Accident Investigation Symposium, 1983, page 8). This period of cooperation and focus makes it possible to ask questions that are not usually asked, gather data not usually gathered, and probe issues not usually open to inquiry.

Not all stakeholders are in the same position relative to the surprising event. Some are closer; others more distant in knowledge, point of view, and in experiencing the consequences. Distance from the work context where an accident occurs appears to alter what is learned from incidents and accidents. Those people who are at the epicenter of high consequence accidents are usually devastated and entirely caught up in the consequences and reactions to failure (Dekker, 2003b). Conversely, people who are far distant from the epicenter are too divorced from the complex context of technical work and accept skeletal descriptions of the events that reach them (the first stories in Cook et al., 1998). As a result, those distant from the technical work area tend to fall back on oversimplified sterile responses.

But near the epicenter are people who have both a detailed understanding of the context of work yet are sufficiently distant from the consequences. Because their technical work corresponds closely to conditions of work at the epicenter, the event has direct relevance. Because they are some distance from the epicenter, their attention is not captured by the need to react to the event itself and they have an opportunity to extract deeper information (they can begin to explore the second stories of Cook et al., 1998), i.e., to learn about how safety is created.

Barriers to Learning

Learning in the aftermath of incidents and accidents is extraordinarily difficult because of the complexity of modern systems. Layers of technical complexity hide the significance of subtle human performance factors. Awareness of hazard and the consequences of overt failure lead to the deployment of (usually successful) strategies and defenses against failure. These efforts create a setting where overt failures only occur when multiple small faults combine. The combination of multiple contributors and hindsight bias makes it easy for reviewers after the fact to identify an individual, group or organization as a culprit and stop. These characteristics of complex systems tend to hide the real characteristics of systems that lead to failures.

When an organization experiences an incident, there are real, tangible and sometimes tragic consequences associated with the event which create barriers to learning:

- The negative consequences are emotional and distressing for all concerned,
- failure generates pressure from different stakeholders to resolve the situation,
- a clear understandable cause and fix helps stakeholders move on from a tragedy, especially when they continue to use or participate in that system,
- managing financial responsibility for ameliorating the consequences and losses from the failure,
- desire for retribution from some stakeholders and processes of defense against punitive actions,
- confronting dissonance and changing concepts and ways of acting is painful and costly in non-economic senses.

In this chapter we present a case study of learning from incidents. Analysis of the case reveals a discounting or distancing process whereby reviewers focus on differences, real and imagined, between the place, people, organization and circumstances where an incident happens and their own context. By focusing on the differences, they see no lessons for their own operation and practices or only narrow well bounded responses. We call this pattern-distancing through differencing.

Examining how this particular organization struggled to recognize and overcome distancing through differencing also provides useful insights on how to support the organizational learning process.

An Incident

A chemical fire occurred during maintenance on a piece of process machinery in the clean room of a large, high technology product manufacturing plant. The fire was detected and automatically extinguished by safety systems that shut off the flow of reactants to the machine.

The reactant involved in the fire was only one of many hazards associated with this expensive machine and the machine was only one of many arranged side by side in a long bay. Operation and maintenance of the machine also involved exposure or potential exposure to thermal, chemical, electrical, radio frequency, and mechanical hazards. Work in this environment was highly proceduralized and the site had repeatedly undergone ISO 9000 certification and review. Both the risks of accident and the high value of the machine and its operation had generated elaborate formal procedures for maintenance and required two workers (buddy system) for most procedures on the machine.

The manufacturer had an extensive safety program that required immediate and high level responses to an incident such as this, even though no personal injury occurred and damage was limited to the machine involved. High level management directed immediate investigations, including detailed debriefings of participants, reviews of corporate history for similar events, and a 'root cause' analysis. Company policy required completion of this activity within a few days and formal, written notification of the event and related findings to all other manufacturing plants in the company. The cost of the incident may have been more than a million dollars (and the plant's score card suffered significantly).

Two things prompted the company to engage outside consultants for a broader review of the accident and its consequences. First, a search for prior similar events in the company files discovered a very similar accident at a manufacturing plant in another country earlier in the year. Second, one of the authors (RIC) recently had been in the plant to study the use of a different machine where operator 'error' seemed prevalent but only with economic consequences. He had identified a systemic trap in this other case and provided some education about how complex systems failed a few weeks earlier. During that visit, he pointed out how other systemic factors could contribute to future incidents that threatened worker safety in addition to economic losses and suggested the need for broader investigations of future events.

Following the incident the authors returned, visited the accident scene, and debriefed the participants in the event and those involved in its investigation. They studied operations involving the machine in which the fire occurred. They also examined the organizational response to this accident and to the prior fire.

Organizational Learning in this Case

The obstacles to learning from failure are nearly as complex and subtle as the circumstances that surround a failure itself. Because accidents always involve multiple contributors, the decision to focus on one or another of the set, and therefore what will be learned, is largely socially determined.

In the incident just described, the formal process of evaluating and responding to the event proceeded along a narrow path. The investigation concentrated on the machine itself, the procedures for maintenance, and the operators who performed the maintenance tasks. For example, they identified the fact the chemical reactant lines were clearly labeled outside the machine but not inside it where the maintenance took place. These local deficiencies were corrected quickly. In a sense, the accident was a 'normal' occurrence in the company; the event was regretted, undesirable, and costly but essentially the sort of thing for which the company's incident procedures had been designed and response teams created. The main findings of this formal, internal investigation were limited to these rather concrete, immediate, local items.

A broader review, conducted in part by outsiders, was based on using the specific incident as a wedge to explore the nature of technical work in context and how workers coped with the significant hazards inherent in the manufacturing process. This analysis yielded a different set of findings regarding both narrow human engineering deficiencies and organizational issues. In addition to the relatively obvious human engineering deficiencies in the machine design discovered by the formal investigation, the event pointed to deeper issues that were relevant to other parts of the process and other potential events.

There were significant limitations in procedures and policies with respect to operations and maintenance of the machine. For example, although there were extensive procedural specifications contained in maintenance 'checklists', the workers had been called on to perform multiple procedures at the same time and had to develop their own task sequencing to manage the combination. Similarly, although the primary purpose of the buddy system was to increase safety by having one worker observe another to detect incipient failures, it was impossible to have an effective buddy system during critical parts of the procedures and parts of this maintenance activity. Some parts of the procedures were so complex that one person had to read the sequence from a computer screen while the other performed the steps. Other steps required the two individuals to stand on opposite sides of the machine to connect or remove equipment, making direct observation impossible.

Surprisingly, the formal process of investigating accidents in the company actually made deeper understanding of accidents and their sources more difficult. The requirement for immediate investigation and reporting contributed to pressure to reach closure quickly and led to a quick superficial study of the incident and its sources. The intense concern for 'safety' had led the company to formally lodge responsibility for safety in a specific group of employees rather than the production and maintenance workers themselves. Treating safety as an abstract goal generated the need for these people as a separate entity within the company. These 'safety people' had highly idealized views of the actual work environment, views uninformed by day to day contact with the realities of clean room work conditions. These views allowed them to conceptualize the accident as flowing from the workers rather than the work situation. They were captivated in their investigation by physical characteristics of the workplace, especially those characteristics that suggested immediate, concrete interventions that could be applied to 'fix' the problems that they thought led to the accident.

In contrast, the operators regarded the incident investigation and proposed countermeasures as derived from views that were largely divorced from the realities of the workplace. They saw the 'safety people' and their work as being irrelevant. They delighted in pointing out, for example, how few of them had any practical experience with working in the clean room. Privately, the workers said that production pressures were of paramount importance in the company. This view was communicated clearly to the workforce by multiple levels of management. Only after accidents, they noted, was safety regarded as a primary goal; during normal operations, safety was always a background issue, in contrast to the primary need to maintain high rates of production. The workers themselves internalized this view. There were significant incentives provided directly to workers to obtain high production and they generally sought high levels of output to earn more money.

During the incident investigation, it was discovered that a very similar incident had occurred at another manufacturing plant in another country earlier in the year – a precursor event or rehearsal from the point of view of this manufacturing facility. Within the company, every incident, including the previous overseas fire, was communicated within the company to safety people and then on to other relevant parties. However, the formal report writing and dissemination about this previous incident had been slow and incomplete, relative to when the second event occurred. Part of the recommendations following from the second incident addressed faster production and circulation of reports (in effect, increasing the pressure to reach closure when investigating incidents).

Interestingly, the relevant people at the plant knew all about the previous incident as soon as it had occurred through more informal communication channels. They had reviewed the incident, noted many features that were different from their plant (non-US location, slightly different model of the same machine, different safety systems to contain fires). The safety people consciously classified the incident as irrelevant to the local setting, and they did not initiate any broader review of hazards in the local plant. Overall they decided the incident "couldn't happen here."

This is an instance of a discounting or distancing process whereby reviewers focus on differences, real and imagined, between the place, people, organization and circumstances where an incident happens and their own context. By focusing on the differences, they see few or no lessons for their own operation and practices.

Notice how speeding up formal notification does nothing to enhance what is learned and does nothing to prevent or mitigate discounting the relevance of the previous incident. The formal review and reports of these incidents focused on their unique features. This made it all the easier for audiences to emphasize what was different and thereby limit the opportunity to learn before they experienced their own incident.

It is important to stress that this was a company taking safety seriously. Within the industry it had an excellent safety record and invested heavily in safety. Its management was highly motivated and its relationships with workers were good, especially because of its strong economic performance that led to high wages and good working conditions. It recognized the need to make a corporate commitment to safety and to respond quickly to safety-related events. Strong pressures to act quickly to 'make it safe' provided incentives to respond immediately to each individual accident. But these demands in turn directed most attention after an accident towards specific countermeasures designed to prevent recurrence of that specific accident. This, in turn, led to the view that accidents were essentially isolated, local phenomena, without wider relevance or significance.

The management of the company was confronted with the fact that the handling of the overseas accident had not been effective in preventing the local one, despite their similarities. They were confronted by the effect of social processes working to isolate accidents and making them seem irrelevant to local operations. The prior fire overseas was noticed but regarded as irrelevant until after the local fire, when it suddenly became critically important information. It was not that the overseas fire was not communicated. Indeed it was observed by management and known even to the local operators. But these local workers regarded the overseas fire not as evidence of a type of hazard that existed in the local workplace but rather as evidence that workers at the other plant were not as skilled, as motivated and as careful as they were after all, they were not Americans (the other plant was in a first world country). The consequence of this view was that no broader implications of the fire overseas were extracted locally after that event.

Interestingly (and ominously) this distancing through differencing that occurred in response to the external, overseas fire, was repeated internally after the local fire. Workers in the same plant, working in the same area in which the fire occurred but on a different shift, attributed the fire to lower skills of the workers on the other shift. (Workers and managers of other parts of the manufacturing process also saw little relevance or potential to learn from the event.) They regarded the workers to whom the accident happened as inattentive and unskilled. Not surprisingly, this meant that they saw the fire as largely irrelevant to their own work. After all, their reasoning went, the fire occurred because the workers to whom it happened were less careful than we are. Despite their beliefs, there was no evidence whatsoever that there were significant differences between workers on different shifts or in different countries (in fact, there was evidence that one of the workers involved was among the better skilled at this plant).

Contributing to this situation was, paradoxically, safety. Over a span of many years, the incidence of accidental fires with this particular chemical and in general had been reduced. But as a side effect of success, personnel's sensitivity to the hazard the chemical presented in the workplace was reduced as well. Interviews with experienced 'old hands' in the industry indicated that such fires were once relatively common. New technical and procedural defenses against these events had reduced their frequency to the point that many operators had no personal experience with a fire. These 'old hands' were almost entirely people now in management positions, far from the clean room floor itself. Those working with the hazardous materials were so young that they had no personal knowledge of these hazards, while those who did have experience were no longer involved in the day to day operations of the clean room.

In contrast with the formal investigation, the more extensive look into the accident that the outside researchers' visit provoked produced different findings. Discussion of the event prompted new observations from within the plant. Two examples may be given. One manager observed that the organization had extensive and refined policies for the handling of the flammable chemical delivery systems (tanks, pipes, valves) that stopped at the entrance to the machine. Different people, policies, and procedures applied to the delivery system. He made an argument for carrying these rules and policies through to the machine itself. This would have required more extensive (and expensive) preparation for maintenance on the machine than was currently the case, but would have eliminated the hazardous chemical from within the machine prior to beginning maintenance. Another engineer suggested that the absence of appropriate labeling on the machine involved with the accident should prompt a larger review of the labeling in all places where this chemical was used or transported.

These two instances are examples of using a specific accident to discover characteristics of the overall system. This kind of reasoning from the specific to the more general is a pronounced departure from the usual approach of narrowly looking for ways to prevent a very specific event in a specific place from occurring or reoccurring.

The chemical fire case reveals the pressures to discount or distance ourselves from incidents and accidents. In this organization, effective by almost all standards, managers, safety officers, and workers took a narrow view of the precursor event. By narrowing in on local, concrete, surface characteristics of the precursor event, the organization limited what could be learned.

Extending or Enhancing the Learning Opportunity

An important question for resilience management is a better understanding of how the window of opportunity for learning can be extended or enhanced following incidents. The above case illustrates one general principle that could be put into action by organizations – do not discard other events because they appear on the surface to be dissimilar. At some level of analysis, all events are unique; while at other levels of analysis, they reveal common patterns.

Promoting means for organizations to look for and consider similarities between their own operation and the organization where an incident occurred could reduce the potential for distancing through differencing. This will require shifting analysis of the case from surface characteristics to deeper patterns and more abstract dimensions (Cook et al., 1998). Each kind of contributor to an event then can guide the search for similarities.

When this process of learning moved past the obstacle of distancing through differencing in this case, the organizational response changed. The organization derived and shared with us a new lesson – safety is a value of an organization, not a commodity to be counted or a priority set among many other goals.

Chapter 21

States of Resilience

Erik Hollnagel
Gunilla Sundström

Introduction

A resilient system, or, organisation is able to withstand the effects of stress and strain and to recover from adverse conditions over long time periods. One way of describing that is to think of a system as being in one of several states, where a state can be defined as 'any well-defined condition or property that can be recognised if it occurs again' (Ashby, 1956, p. 25). The set of possible states constitutes the so-called state space, and each state has a set of associated conditions that specify whether the system changes to another state. The state space description is ubiquitous and can be applied to almost any type of system. The simplest possible case is, of course, a system that can be in either of two states or conditions. This applies to an uncomplicated device such as an electrical switch, which can be either 'on' or 'off', but also to a complex system such as a national economy that can be either growing or receding. A car engine might be described by means of three states, namely 'stopped', 'idle', or 'running'. Even a mighty nation can – from one perspective – be characterised as being in one of five states according to the threat condition, such as 'green: low', 'blue: guarded', 'yellow: elevated', 'orange: high', and 'red: severe'.

In Chapter 15, a business system was described as being in a healthy, an unhealthy or a catastrophic state with the possible transitions being shown in Figure 15.2. Note that the state space in this case, as well as the case of the five 'threat' states, is one-dimensional, which means that it is only possible to make transitions between adjacent states. To take a slightly more complex example, consider the following description of the state space of a power plant. Here there are six different states, as well as the possible transitions between them. Without going into any details, the diagram in Figure 21.1 shows that while some transitions are bi-directional, e.g., between 'normal operation' and 'disturbance', others are not reversible but require that the system passes through one or more intermediate states. For instance, once the system has gone to a state of 'emergency' it can only return to 'normal operation' via an 'idle' state.

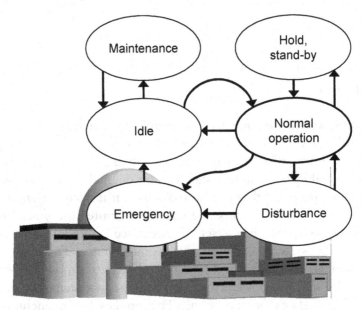

Figure 21.1: State-space diagram for power production

Resilience and State-space Transitions

The terminology used in Figure 21.1 is typical for systems that produces something physical, matter or energy, where the focus is on operations rather than services. The same principles can, however, be used to characterise various kinds of organisations, such as financial services firms, government departments, regulatory entities, rescue organisations, etc. Even if we disregard the extreme periods of a system's life, hence exclude embryonic and decaying organisations, it is easily possible to recognise characteristic states that are more detailed than the triad used in Chapter 15 (healthy, unhealthy, catastrophic), cf. Figure 21.2. There must clearly always be a state of normal functioning, where the system provides or produces what it is intended to do in a reliable and, if required, in a profitable manner. There will usually also be a state of regular reduced functioning, for instance during night time or holidays, as well as a state of irregular reduced functioning due to a lack of internal resources (illness among staff, equipment failures, industrial actions, and the like). The transitions between normal and regular reduced functioning are scheduled, whereas the transitions between normal and irregular reduced functioning usually are unexpected and represent a temporary loss of control. Since even the most sanguine managers know that things can go wrong, a state of irregular reduced functioning has usually also been anticipated and adequate recovery functions are therefore hopefully in place.

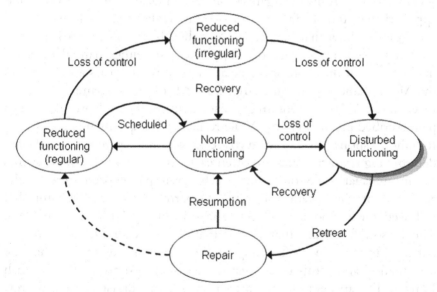

Figure 21.2: State-space diagram for service organisations

There will, unfortunately, also always be a state of disturbed functioning, corresponding to the unhealthy or even catastrophic states of Chapter 15. Figure 21.2 only shows one state of disturbed functioning, but there may of course be several. Indeed, it may be the mark of a resilient organisation that it has a number of different modes of functioning whenever a disturbance happens. The transition to a state of disturbed functioning clearly represents a loss of control. The return to a state of normal functioning may either be through a direct recovery, or in the case of severe disturbances via a state of repair. In extreme cases there is, of course, the possibility that the system ceases to exist after a severe disturbance, as the case of Barings PLC demonstrated. The state transitions described here are representative but not complete; under unusual circumstances other transitions may become possible.

The Sumatra Tsunami Disaster

To illustrate how a state space description can be used to understand the nature of resilience, consider how an organisation responds to a severe event. A resilient organisation should clearly be able to respond appropriately and as fast as possible in order not to lose control. Examples of how that happens – or rather, how it does not happen – are often seen in the case of major disasters, either natural or man-made. Some of the more spectacular cases in recent years are the fire in the Mont Blanc tunnel on March 26 1999 or the tsunami in Asia on December 26, 2004. (One might also mention the collapse of Barings plc described in Chapter 15.) While resilience is a quality that is essential for the response to all disturbances, be they large or small, the determining characteristics are often easier to note in the case of events of an unusual scale or severity. The example chosen here is the response of the Swedish Foreign Department to the tsunami that followed the earthquake off Sumatra on Sunday December 26 2004. The reason for using that is not that the Foreign Department was particularly inept in its handling of the situation, but merely that the information about this case is easily accessible. (Although the Swedish Foreign Department in the aftermath of the disaster was severely criticised at home, foreign departments in several other European countries got a similar treatment from their national press.)

A tsunami following a major earthquake is the type of event that requires a fast and appropriate response. It is, however, also a type of event that is very unusual and one that was not specifically foreseen in the procedures of the Foreign Department. Although the Foreign Department received information about the disaster around 5 o'clock in the morning of December 26, it did not begin to respond effectively until the next day. In the aftermath the Secretary of Foreign Affairs openly admitted that no one at the Foreign Department understood the scope of the disaster, and that she herself did not even know where Phuket was. (As it happened, the Phuket area was a popular destination for Swedish travellers. Several thousand visitors from Sweden were there during the 2004 Christmas holidays and 544 of them either died or went missing after the tsunami.) Partly because of that, and partly because of the holidays, the Foreign Department was not fully staffed and therefore responded very slowly. In contrast to that, both the charter companies that were responsible for the tourists in Thailand, the Swedish police, and the Swedish Rescue Services Agency went into a state of high alert within hours. Indeed, one of the travel companies had posted information about the tsunami around 7 o'clock in the morning, i.e., about two hours after they received the information from Thailand.

In terms of the concepts described above, the Foreign Department remained for too long in the regular reduced functioning state (probably somewhat due to the Christmas holiday season). There were no clear routines or procedures for what to do, and because of that offers of assistance, e.g., from the police, were turned down. The situation was further aggravated by the department's utter failure to realise what was going on, despite the fact that the TV channels had extra newscasts throughout the day. When finally the magnitude of the disaster was realised, there was little capability to do anything.

If we consider the experiences from this disaster in terms of a state transition description, resilience can be seen to require three things: an ability to recognise that conditions have changed and that there is a need to respond, a set of transition rules (procedures or routines that allow the system to transition from one state to the other by changing its mode of operation), and a readiness or capability for getting on with the tasks at hand once the new state has been entered. To that may be added a fourth, namely the ability to maintain normal operations throughout. This is, however, not an absolute requirement, but depends on the type of organisation. For a foreign department it is clearly desirable to maintain normal operations even when an emergency arises; for the charter companies it was possible to shift the whole organisation into a new mode and suspend all normal functions as long as it lasted.

Conclusions

It is a common view that delays in changes or transitions often are due to fossilised organisational structures or inadequate and/or conflicting policies. Many organisations develop a cumbersome bureaucracy with built-in social conflicts where people at various levels are unable to, or incapable of, making the appropriate decisions. Decisions cannot be made on the spot but have to be passed through organisational layers until they reach a proper level, often at the top of the hierarchy. The resulting decisions and their translation into actions then have to go through the reverse process – although this often is considerably faster, partly because the decision resulted in a state change. Decision makers may with hindsight often state that they did not fully appreciate the seriousness of the situation, or that procedures were not in place for the required action. Yet from the point of view of resilience engineering it simply means that the organisation was not sufficiently resilient.

A resilient organisation must, however, not only be able to change from one state to a more appropriate one in time, but also be able to return to normal functioning when the alerting or unusual conditions are over. This does not necessarily mean that it should go back to what were normal procedures before the events, since the world may have changed. But it means that it should be able to resume durable and sustainable performance, in the sense of attaining at least the same quantity and quality of whatever it produces, as it did before the disrupting events.

The ability to rebound or recover again requires that the cessation of the abnormal state is detected, that there are proper procedures for returning or reverting to 'normal', and that the functions and capabilities required for a normal – or a revised normal – operation are in place. In cases when the emergency and normal operations run in parallel, returning to normal conditions is fairly simple. The normal system state must, however, allow the organisation somehow to absorb the resources that are released when the emergency operation comes to an end, possibly by keeping them idle.

In summary, the state transition analogy offers a number of concrete (potential) measures and procedures for the resilience engineering toolbox.

- First, there is the issue of being able to recognise the changed situation, i.e., the triggering conditions for the transition. Here a host of organisational, psychological, and social factors may work against that – ignorance, biases and vested interests – in addition to more 'simple' technical problems such as lack of sufficient data. The associated measure is the organisation's ability to diagnose, assess or understand the situation and anticipate the consequences. This ability is typically documented by rules and procedures, time and resources allocated to this task, etc.

- Second, there is the issue of procedures or routines (i.e., change management processes) for the transition. Do people in the organisation know what to do? Are command lines defined and generally understood? Is it known who has responsibility, and whether the normal roles change? Is there a script or scenario, a set of guidelines or even a set of instructions that can be applied?

- Third, does the 'new' (i.e., receiving) state provide the resources and capabilities needed for the desired functionality? Is there perhaps a back-up organisation ready? Do physical and communication facilities exist? Are required supplies and resources available? Are there sufficient human resources, human capital and experience? The associated measure is whether the organisation is ready to respond when a new state is entered, i.e., whether strategies, policies, procedures, resources, capabilities, and so on exist. Are there regular exercises? Do people know their roles and responsibilities?

- A possible fourth issue is whether the transition is directly reversible in principle and/or in practice, or whether the organisation must pass through a repair or reconstruction state. At issue here is whether the organisation is able to stop or dismantle and/or significantly re-engineer an operation, i.e., to return from an abnormal to a normal state. Related to that is the issue of whether the cessation of a condition can be detected when it happens, but not too early.

In relation to the first step, recognising the changed situation, one additional issue is how obvious the change must be. This relates to the discussion elsewhere (Chapter 1) about being reactive or proactive. In the reactive case, which is the most common, a time lag necessarily exists between the change and its detection, but a resilient organisation will ensure that it is as small as possible. In the proactive case, the time lag is reduced or perhaps even eliminated. Yet even in this case the response is based on some cues; and the uncertainty of the outcome is increased in the sense that the action(s) taken may be inappropriate.

Epilogue: Resilience Engineering Precepts

Erik Hollnagel
David D. Woods

Safety is Not a System Property

One of the recurrent themes of this book is that safety is something a system or an organisation *does*, rather than something a system or an organisation *has*. In other words, it is not a system property that, once having been put in place, will remain. It is rather a characteristic of how a system performs. This creates the dilemma that safety is shown more by the absence of certain events – namely accidents – than by the presence of something. Indeed, the occurrence of an unwanted event need not mean that safety as such has failed, but could equally well be due to the fact that safety is never complete or absolute.

In consequence of this, resilience engineering abandons the search for safety as a property, whether defined through adherence to standard rules, in error taxonomies, or in 'human error' counts. By doing so it acknowledges the danger of the reification fallacy, i.e., the tendency to convert a complex process or abstract concept into a single entity or thing in itself (Gould, 1981, p. 24). Seeing resilience as a quality of functioning has two important consequences.

- We can only measure the *potential* for resilience but not resilience itself. Safety has often been expressed by means of reliability, measured as the probability that a given function or component would fail under specific circumstances. It is, however, not enough that systems are reliable and that the probability of failure is below a certain value (cf. Chapter 16); they must also be resilient and have the ability to recover from irregular variations, disruptions and degradation of expected working conditions.

- Resilience cannot be engineered simply by introducing more procedures, safeguards, and barriers. Resilience engineering instead requires a continuous monitoring of system performance, of how things are done. In this respect resilience is tantamount to coping with complexity (Hollnagel & Woods, 2005), and to the ability to retain control.

Resilience as a Form of Control

A system is in control if it is able to minimise or eliminate unwanted variability, either in its own performance, in the environment, or in both. The link between loss of control and the occurrence of unexpected events is so tight that a preponderance of the latter in practice is a signature of the former. Unexpected events are therefore often seen as a *consequence* of lost control. The loss of control is nevertheless not a necessary condition for unexpected events to occur. They may be due to other factors, causes and developments outside the boundaries of the system.

An unexpected event can also be a *precipitating* factor for loss of control and in this respect the relation to resilience is interesting. Knowing that control has been lost is of less value than knowing *when* control is going to be lost, i.e., when unexpected events are likely. In fact, according to the definition of resilience (Chapter 1), the fundamental characteristic of a resilient organisation is that it does not lose control of what it does, but is able to continue and rebound (Chapter 13).

In order to be in control it is necessary to know what has happened (the past), what happens (the present) and what may happen (the future), as well as knowing what to do and having the required resources to do it. If we consider joint cognitive systems in general, ranging from single individuals interacting with simple machines, such as a driver in a car, to groups engaged in complex collaborative undertakings, such as a team of doctors and nurses in the operating room, it soon becomes evident that a number of common conditions characterise how well they perform and when and how they lose control, regardless of domains. These conditions are lack of *time*, lack of *knowledge*, lack of *competence*, and lack of *resources* (Hollnagel & Woods, 2005, pp. 75-78).

Lack of time may come about for a number of reasons such as degraded functionality, inadequate or overoptimistic planning, undue demands from higher echelons or from the outside, etc. Lack of time is, however, quite often a consequence of lack of foresight since that pushes the system into a mode of reactive responding. Knowing what happens and being able to respond are not by themselves sufficient to ensure control, since a system without anticipation is limited to purely reactive behaviour. That inevitably incurs a loss of time, both because the response must come after the fact and therefore be compensatory, and because the resources to respond may not always be ready when needed but first have to be marshalled. In consequence of that, a system confined to rely on feedback alone will in most cases sooner or later fall behind the pace of events and therefore lose control.

Knowledge is obviously important both for knowing what to expect (anticipation) and for knowing what to look for or where to focus next (attention, perception). The encapsulated experience is sometimes referred to as the system's 'model of the world' and must as such be dynamic rather than static. Knowledge is, however, more than just experience but also comprises the ability to go beyond experience, to expect the unexpected and to look for more than just the obvious. This ability, technically known as requisite imagination (Westrum, 1991; Adamski & Westrum, 2003), is a *sine qua non* for resilience.

Competence and resources are both important for the system's ability to respond rationally.[1] The competence refers to knowing what to do and knowing how to do it, whereas the resources refer to the ability to do it. That the latter are essential is obvious from the fact that control is easily lost if the resources needed to implement the intended response are missing. This is, for instance, a common condition in the face of natural disasters such as wildfires, earthquakes, and pandemics.

Figure E.1 illustrates three qualities that a system must have to be able to remain in control, and therefore to be resilient, with time as a fourth, dependent quality. The three main qualities are not linked in the sense that anticipation precedes attention, which in turn precedes response. Although this ordering in some sense will be present for any

---

[1]   Being rational is not used in the traditional, normative meaning, but rather as the quality of being anti-entropic, cf. Hollnagel (2005).

specific instance that is described or analysed, the whole point about resilience is that these qualities must be exercised continuously. The system must constantly be watchful and prepared to respond. Additionally, it must constantly update its knowledge, competence and resources by learning from successes and failures – its own as well as those of others.

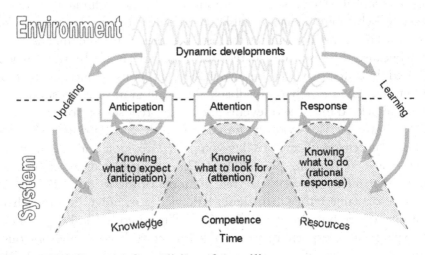

Figure E.1: Required qualities of a resilient system

It is interesting to note that Diamond (2005) in his book on how societies collapse and go under, identifies three 'stops on the road to failure' (p. 419). These are: (1) the failure to anticipate a problem before it has arrived, (2) the failing to perceive a problem that has actually arrived, and (3) the failure to attempt to solve a problem once it has been perceived (rational bad behaviour). A society that collapses is arguably an extreme case of lack of resilience, yet it is probably no coincidence that we find the positive version of exactly the same characteristics in the general descriptions of what a system – or even an individual – needs to remain in control. A resilient system must have the ability to anticipate, perceive, and respond. Resilience engineering must therefore address the principles and methods by which these qualities can be brought about.

Readiness for Action

It is a depressing fact that examples of system failures are never hard to find. One such case, which fortunately left no one harmed, occurred during the editing of this book. As everybody remembers, a magnitude 9.3 earthquake occurred in the morning of December 26, 2004, about 240 kilometers south of Sumatra. This earthquake triggered a tsunami that swept across the Indian Ocean killing at least 280,000 people. One predictable consequence of this most tragic disaster was that coastal regions around the world became acutely aware of the tsunami risk and therefore of the need to implement well-functioning early warning systems. In these cases there is little doubt about what to expect, what to look for, and what to do. So when a magnitude 7.2 earthquake occurred on June 14, 2005, about 140 kilometres off the town of Eureka in California, the tsunami warning system was ready and went into action.

As it happened, not one but two tsunami warning centres reacted. The first warning, covering the US and Canadian west coast, came from a centre in Alaska. Three minutes later a second warning was issued by a centre in Hawaii. The second warning said that there was no risk of tsunami, but excluded the west coast north of California from the warning. Rescue workers, missing this small but significant detail, were predictably confused (Biever & Hecht, 2005, p. 24).

Tsunami warnings are broadcast via radio by the US National Oceanic and Atmospheric Administration (NOAA). Unfortunately, some locations cannot receive the NOAA radio signals because they are blocked by mountains. They are therefore contacted by phone from Seattle. On the day in question, however, a phone line was down so that the message did not get through, effectively leaving some areas without warning. This glitch was, however, not noticed. As it happened, the earthquake was of a type that could not give rise to a tsunami, and the warning was cancelled after one hour.

This example illustrates a system that was not resilient, despite being able to detect the risk in time. While precautions had been made and procedures put in place, there was no awareness of whether they actually worked and no understanding of what the actual conditions are. The specific shortcoming was one of communication, issuing inconsistent warnings and lacking feedback, and the consequence was a partial lack of readiness to respond. Using the terminology proposed in

Chapter 21, the communication failure meant that some districts did not go into the required state of high alert, as a preparation for an evacuation. While the tsunami warning system was designed to look for specific factors in the environment, it was not designed to look at itself, to ensure that the 'internal' functions worked. The system was designed to be safe by means of all the technology and procedures that were put in place, but it was not designed to be resilient.

Why Things Go Wrong

It is a universal experience that things sooner or later will go wrong,[2] and fields such as risk analysis and human reliability assessment have developed a plethora of method to help us predict when and how it may happen. From the point of view of resilience engineering it is, however, at least as important to understand *why* things go wrong. One expression of this is found in the several accident theories that have been proposed over the years (e.g., Hollnagel, 2004), not least the many theories of 'human error' and organisational failure. Most such efforts have been engrossed with the problems found in technical or socio-technical systems. Yet there have also been attempts to look at the larger issues, most notably Robert Merton's lucid analysis of why social actions often have unanticipated consequences (Merton, 1936).

It is almost trivial to say that we need a model, or a frame of reference, to be able to understand issues such as safety and resilience and to think about how safety can be ensured, maintained, and improved. A model helps us to determine which information to look for and brings some kind of order into chaos by providing the means by which relationships can be explained. This obviously applies not only to industrial safety, but to every human endeavour and industry. To do so, the model must in practice fulfil two requirements. First, that

---

[2]   This is often expressed in terms of Murphy's law, the common version of which is that 'everything that can go wrong, will'. A much earlier version is Spode's law, which says that 'if something can go wrong, it will.' It is named after the English potter Josiah Spode (1733-97) who became famous for perfecting the transfer printing process and for developing fine bone china – but presumably not without many failed attempts on the way.

it provides an explanation or brings about an understanding of an event such that effective mitigating actions can be devised. Second, that it can be used with a reasonable investment of effort – intellectual effort, as well as time and resources. A model that is cumbersome and costly to use will, from an academic point of view, from the very start be at a disadvantage, even if it provides a better explanation.[3] The trick is therefore to find a model that at the same time is so simple that it can be used without engendering problems or requiring too much specialised knowledge, yet powerful enough to go beneath the often deceptive surface descriptions.

The problem with any powerful model is that it very quickly becomes 'second nature', which means that we no longer realise the simplifications it embodies. This should, however, not lead to the conclusions that we must give up on models and try to describe reality as it really is, since that is a philosophically naïve notion. The consequence is rather that we should acknowledge the simplifications that the models bring, and carefully weigh advantages against disadvantages so that a choice of model is made knowingly.

Several models have been mentioned in the chapters in this book. The most important models in the past have been the Domino model and the Swiss cheese model (Chapter 1). Both are easy to comprehend and have been immensely helpful in improving the understanding of accidents. Yet their simplicity also means that some aspects cannot be easily described – or described at all, and that explanations in terms of the models therefore may be incomplete. (Strictly speaking, the two models are metaphors rather than models. In one, accidents are likened to a row of dominoes falling, and in the other, to harmful influences passing through a series of holes aligned.)

In the case of the Domino model, it is clear that the real world has no domino pieces waiting to fall. There may be precariously poised

---

[3]   'Better' is, of course, a rather dangerous term to use since it implies that some objective criterion or standard is available. Although there is no truth to be used as a point of reference, it is possible to show that one explanation – under given conditions – may be better than another, e.g., in providing more effective countermeasures. Changes are, however, never contemplated *sub specie aeternatis* but are always subject to often very mundane or pecuniary considerations.

systems or subsystems that suddenly may change from a normal to an abnormal state, but that transition is rarely as simple as a domino falling. Likewise, the linking or coupling between dominoes is never as simple as the model shows. Similarly, the Swiss cheese model does not suggest that we should look for slices of cheeses or holes, or measure the size of holes or movements of slices of cheese. The Swiss cheese model rather serves to emphasise the importance of latent conditions and illustrate how these in combination with active failures may lead to accidents.

The Domino and Swiss cheese models are useful to explain the abrupt, unexpected onset of accidents, but have problems in accounting for the gradual loss of safety that may also lead to accidents. In order to overcome this problem, a model of 'drift to danger' has been used, for example in Chapter 3. Although the metaphor of drift introduces an important dynamic aspect, it should not be taken literally or as a model, for the following reasons:

- Since the boundaries or margins only exist in a metaphorical sense or perhaps as emergent descriptions (Cook & Rasmussen, 2005), there is really no way in which an organisation can 'sail close' to an area of danger, nor ways in which the 'distance' can be measured. 'Drift' then only refers to how a series individual actions or decisions have larger, combined and longer term impacts on system properties that are missed or underappreciated.

- The metaphor itself oversimplifies the situation by referring to the organisation as a whole. There is ample practical experience to show that some parts of an organisation may be safe while others may be unsafe. In other words, parts of the organisation may 'drift' in different directions. The safety of the organisation can furthermore not be derived from a linear combination of the parts, but rather depends on the ways in which they are coupled and how coordination across these parts is fragmented or synchronized (cf. Perrow, 1984). This is also the reason why accidents in a very fundamental sense are non-linear phenomena.

- Finally, there are no external forces that, like the wind, push an organisation in some direction, or allow the 'captain' to steer it clear of danger. What happens is rather that choices and decisions made during daily work may have long-term consequences that are not considered at the time. There can be many reasons for this, such as the lack of proper 'conceptual' tools or a shortage of time.

It is inevitable that organisational practices change as part of daily work, one simple reason being that the environment is partly unpredictable, changing, or semi-erratic. Such changes are needed either for purposes of safety or efficiency, though mostly the latter. Indeed, the most important factor is probably the need to gain time in order to prevent control from being lost, as described by the efficiency-thoroughness trade-off (ETTO; Hollnagel, 2004). There is never enough time to be sufficiently thorough; finishing an activity in time may be important for other actions or events, which in turn cannot be postponed because yet others depend on them, etc. The reality of this tight coupling is probably best illustrated by the type of industrial action that consists in 'working to rule.' This also provides a powerful demonstration of how important the everyday trade-offs and shortcuts are for the normal functioning of a system.

Changed practices to improve efficiency often have long-term consequences that affect safety, although for one reason or another they are disregarded when the changes are made. These consequences are usually both latent and have latency and therefore only show themselves after a while. Drift is therefore nothing more than an accumulated effect of latent consequences, which in turn result from the trade-off or sacrificing decisions that are required to keep the system running.

A Constant Sense of Unease

Sacrificing decisions take place on both the individual and the organisation levels – and even on the level of societies. While they are necessary to cope with a partly unpredictable environment, they constitute a source of risk when they become entrenched in institutional or organisational norms. When trade-offs and sacrificing decisions become habitual, they are usually forgotten. Being alert or critical incurs a cost that no individual or organisation can sustain permanently and is therefore only used when necessary. Norms *qua* norms are for that reason rarely challenged. Yet it is important for resilience that norms remain conspicuous, not in the sense that they must constantly be scrutinised and revised, but in the sense that their existence is not forgotten and their assumptions take for granted.

Resilience requires a constant sense of unease that prevents complacency. It requires a realistic sense of abilities, of 'where we are'. It requires knowledge of what has happened, what happens, and what will happen, as well as of what to do. A resilient system must be proactive; flexible; adaptive; and prepared. It must be aware of the impact of actions, as well as of the failure to take action.

Precepts

The purpose of this book has been to propose resilience engineering as a step forward from traditional safety engineering techniques – such as those developed in risk analysis and probabilistic safety assessment (PSA). Rather than try to force adaptive processes and organisational factors into these families of measures and methods, resilience engineering recognises the need to study safety as a process, provide new measures, new ways to monitor systems, and new ways to intervene to improve safety. Thinking in terms of resilience shifts inquiry to the nature of the 'surprises' or types of variability that challenge control.

- If 'surprises' are seen as disturbances, or disrupting events, which challenge the proper functioning of a process, then inquiry centres on how to keep a process under control in the face of such disrupting events, specifically on how to ensure that people do not exceed given 'limits.'
- If 'surprises' are seen as uncertainty about the future, then inquiry centres on developing ways to improve the ability to anticipate and respond when so challenged.
- If 'surprises' are seen as recognition of the need constantly to update definitions of the difference between success and failure, then inquiry centres on the kinds of variations which our systems should be able to handle and ways constantly to test the system's ability to handle these classes of variations.
- If 'surprises' are seen as recognition that models and plans are likely to be incomplete or wrong, despite our best efforts, then inquiry

centres on the search for the boundaries of our assessments in order to learn and revise.

Resilience engineering entails a shift from an over-reliance on analysis techniques to adaptive and co-adaptive models and measures as the basis for safety management. Just as it acknowledges and tries to avoid the risks of reification (cf. above), it also acknowledges and tries to avoid the risks of oversimplifications, such as:

- working from static snapshots, rather than recognising that safety emerges from dynamic processes;
- looking for separable or independent factors, rather than examining the interactions across factors; and
- modelling accidents as chains of causality, rather than as the result of tight couplings and functional resonance.

It is fundamental for resilience engineering to monitor and learn from the gap between work as imagined and work as practised. Anything that obscures this gap will make it impossible for the organisation to calibrate its understanding or model of itself and thereby undermine processes of learning and improvement. Understanding what produces the gap can drive learning and improvement and prevent dependence on local workarounds or conformity with distant policies. There was universal agreement across the symposium attendees that previous research supports the above as a critical first principle. The practical problem is how to monitor this gap and how to channel what is learned into organisational practice.

The Way Ahead

This book boldly asserts that sufficient progress has been made on resilience as an alternative safety management paradigm to begin to deploy that knowledge in the form of engineering management techniques. The essential constituents of resilience engineering are already at hand. Since the beginning of the 1990s there has been a growing evolution of the principles for organisational resilience and in the understanding of the factors that determine human and organisational performance. As a result, there is an appreciable basis for how to incorporate human and organisational risk in life cycle systems engineering tools and how to build knowledge management tools that proactively capture how human and organisational factors affect risk.

While additional studies can continue to document the role played by adaptive processes for how safety is created in complex systems, this book marks the beginning of a transition in resilience engineering from research questions to engineering management tools. Such tools are needed to improve the effectiveness and safety of organisations confronted by high hazard and high performance demands. In particular, we believe that further advances in the resilience paradigm should occur through deploying the new measures and techniques in partnership with management for actual hazardous processes. Such projects will have the dual goals of simultaneously advancing the research base on resilience and tuning practical measurement and management tools to function more effectively in actual organisation decision-making.

Appendix

Symposium Participants

René Amalberti is doctor in Medicine (Marseille, 77), Professor of Physiology and Ergonomics (Paris, 95), and a PhD in Cognitive Psychology (Paris 92). He joined the French Air Force in 1977, graduated in aerospace medicine, and is presently head of the cognitive science department at IMASSA (Airforce Aerospace Medical Research Institute).

From 1980 to 1992, he was involved in four major research programs as: (i) developer of the French Electronic co-pilot for fighter aircraft, (ii) developer of an Intelligent-Training support system at Airbus, (iii) team member of the EC Research project Model of Human Activity at Work (MOHAWC), and (iv) co-developer of the first Air-France Crew Resource Management course.

In late 1992 he was detached half-time from the military to the French ministry of transportation, to lead the human factors for Civil Aviation in France, and in 1993 became the first chairman of JAA human factors steering committee, a position he held until late 1999. He has continued as a resource person for civil aviation authorities (HF certification A380, A400M), and manages safety research in various areas such as patient safety (co-writer of the accreditation reference document at the High Agency for Healthcare), road safety (chairman national research program), and the chemical industry (vice chairman of the new National Foundation for Industrial Risk).

René Amalberti has published and co-edited over 100 papers, chapters, and books, most of them on the safety and the emerging theory of Ecological safety.

Lars Axelsson is a Human Factors Specialist at the Swedish Nuclear Power Inspectorate (SKI) in Stockholm, Sweden. He is a specialist in organization and management issues and also works as a consultant for the aviation industry, mostly with pilot assessment and selection, and flight crew training.

Richard I. Cook (MD) is Associate Professor in the Department of Anesthesia and Critical Care at the University of Chicago. His current research interests include the study of human error, the role of technology in human expert performance, and patient safety. Dr. Cook is a founding member of the National Patient Safety Foundation and sits on the Foundation's Board. He is internationally recognized as a leading expert on medical accidents, complex system failures, and human performance at the sharp end of these systems. He has investigated a variety of problems in such diverse areas as urban mass transportation, semiconductor manufacturing, and military software systems. He is often a consultant for not-for-profit organizations, government agencies, and academic groups. Dr. Cook's most often cited publications are 'Gaps in the Continuity of Patient Care and Progress in Patient Safety,' 'Operating at the Sharp End: The Complexity of Human Error,' 'Adapting to New Technology in the Operating Room,' and the report 'A Tale of Two Stories: Contrasting Views of Patient Safety.'

Vinh N. Dang has a PhD in Nuclear Engineering from the Massachusetts Institute of Technology (MIT). In 1994, he joined the Paul Scherrer Institute (PSI) in Switzerland, where he leads research and regulatory support activities related to Human Reliability Analysis (HRA). His research interests include the analysis and quantification of personnel decision-making in accident scenarios, in the frame of Probabilistic Safety Assessments (PSAs); the modeling of crew performance in joint human-system simulations, aimed at developing a dynamic risk assessment methodology to complement PSA; and the development of PSA for applications outside the domain of nuclear power plants. In this area, an on-going PSA study for PSI's Proton Therapy Facility is a notable case combining elements of (medical) physics and the healthcare sector. Currently, he leads a task force on the collection and exchange of data for HRA research and applications, in the Working Group on Risk Assessment of the OECD Nuclear Energy Agency. Recently, he contributed HRA expertise and drew conclusions and recommendations for an IAEA Co-Ordinated Research Project (2001-2003) dealing with exploratory applications of PSA for radiation sources. He also served as Scientific Secretary for the 7th Probabilistic Safety Assessment and Management Conference (PSAM7/ESREL04, Berlin, June 2004).

Sidney Dekker is Professor of Human Factors and Aviation Safety, and Director of Research at Lund University, School of Aviation, Sweden. He received an M.A. in organizational psychology from the University of Nijmegen and an M.A. in experimental psychology from Leiden University, both in the Netherlands. He gained his Ph.D. in Cognitive Systems Engineering from The Ohio State University, USA. He has previously worked for the Public Transport Cooperation in Melbourne, Australia; the Massey University School of Aviation, New Zealand, British Aerospace, UK, and has been a Senior Fellow at Nanyang Technological University in Singapore. His specialties and research interests are system safety, human error, reactions to failure and criminalization, and organizational resilience. He has some experience as a pilot, type trained on the DC-9 and Airbus A340. His books include 'The Field Guide to Human Error Investigations' (2002) and 'Ten Questions About Human Error: A New View of Human Factors and System Safety' (2005).

Arthur Dijkstra is a captain in KLM Royal Dutch Airlines on a Boeing 777 as well as a flight safety investigator. From 1982 to 1984 he completed the Dutch Civil Aviation flying school and started in 1985 as First Officer at the McDonnell Douglas DC 9, followed by Boeing B747-300 and B747-400. On this aircraft he started as a flight instructor. He later became captain on Airbus A310, followed by the Boeing B767 and Boeing B777. From 1998 he was also (Senior) Type Rating examiner, which includes pilot and instructor training and examination. In 2005 he started as flight safety investigator in the flight safety department of KLM. From 2003 to June 2005 Arthur Dijkstra completed a MSc study in Human Factors at Linköping University, Sweden. In 2005 he started a PhD study in the field of Resilience Engineering at the section of Safety Science at the University of Technology in Delft in the Netherlands sponsored by the KLM.

Rhona Flin (BSc, PhD Psychology) is Professor of Applied Psychology in the School of Psychology at the University of Aberdeen. She is a Chartered Psychologist, a Fellow of the British Psychological Society and a Fellow of the Royal Society of Edinburgh. Rhona Flin directs a team of psychologists working with high reliability industries (especially the energy sector) on research and consultancy projects concerned with the management of safety and emergency response. The group has worked on projects relating to aviation safety (funded by the EC and CAA), leadership and safety in offshore management (funded by HSE and the oil industry), health management and safety in the offshore oil industry (HSE), team skills and emergency management (funded by the nuclear industry). She is currently studying anaesthetists' and surgeons' non-technical skills (funded by NES/RCS, Edinburgh) and safety climate in hospitals (funded by GUHT). Her recent books include 'Incident Command: Tales from the Hot Seat' (Ed. & Arbuthnot, Ashgate, 2002); 'Sitting in the Hot Seat: Leaders and Teams for Critical Incident Management' (Wiley, 1996), 'Managing the Offshore Installation Workforce' (Ed. & Slaven, PennWell Books, 1996) and 'Decision Making under Stress' (Ed. & Salas, Strub & Martin, Ashgate, 1997). See also www.abdn.ac.uk/iprc.

Yushi Fujita is a Specialist in the field of human – machine interfaces, individual and organizational characteristics, human cognitive modelling and human reliability analysis, who has over twenty-five years of experience mainly in the nuclear industry. He is currently a managing director at Technova Inc., a research organization in Japan. He received a PhD from the University of Tokyo in 1992.

Andrew Hale has been Professor of Safety Science at the Delft University of Technology since 1984. He worked in London and at Aston University and pioneered the field of occupational accident research, particularly as an occupational psychologist specialising in human factors in safety, safety management and regulation. His early research was centred on individual behaviour and training in the face of danger. Since moving to the Netherlands he has focused more on the modelling and measurement of safety management and culture and on regulation of safety. He has conducted studies in the manufacturing, process and construction industries and more recently in air, road and rail transport.

He is chairman of the Scientific Advisory Board of the Dutch Traffic Safety Institute, member of the Safety Advisory Committee on Schiphol Airport, member of the Executive Board of the Dutch Association of Safety Engineers and editor of *Safety Science*.

The department at Delft has three main research lines: the incorporation of risk as a criterion in the design process of new technology, processes and products; the development, assessment and improvement of risk management systems; and the development and evaluation of tools for risk assessment and modelling and for learning from incidents and accidents.

Erik Hollnagel is since mid-2006 Industrial Safety Chair at Ecole des Mines de Paris, after having been Professor of Human – Machine Interaction at Linköping University (S) since 1999. He is an internationally recognised expert in the fields of system safety, accident analysis, cognitive systems engineering, cognitive ergonomics and intelligent human – machine systems. He has worked at universities, research institutes and private companies in several countries since 1971, and has experience from applications in nuclear power, aerospace, software engineering, healthcare, and road vehicles. He has published widely including ten books, the most recent titles being 'Joint Cognitive Systems: Foundations of Cognitive Systems Engineering' (Taylor & Francis, 2005; co-authored with D. D. Woods) and 'Barriers and Accident Prevention' (Ashgate, 2004), and is, together with Pietro C. Cacciabue, Editor-in-Chief of the international journal of *Cognition, Technology & Work*.

Nancy G. Leveson is Professor of Aeronautics and Astronautics and of Engineering Systems at MIT and Director of the Complex Systems Research Lab. She has over 25 years experience working on system safety, including cross-disciplinary research encompassing engineering, cognitive psychology, and organizational sociology. She has published almost 200 papers on system safety, human – computer interaction, software engineering, and system engineering. She has also published a book entitled *Safeware* (Addison-Wesley, 1995). Dr. Leveson is a member of the National Academy of Engineering and has received numerous awards for her research including the 1999 ACM Allen Newell Award for outstanding computer science research, the 1995 AIAA Information Systems Award for 'developing the field of software safety and for promoting responsible software and system engineering practices where life and property are at stake.' In 2004 she received the ACM Sigsoft Outstanding Research Award. Her research results are currently being used by aerospace, automotive, nuclear power, and defense companies and government agencies. In 1995, Dr. Leveson and some of her former students started a company, Safeware Engineering

Corporation, that creates specification and system safety engineering tools and consults on safety in a wide variety of industries.

Nick McDonald is Senior Lecturer in the Department of Psychology, Trinity College Dublin. He chairs the Aerospace Psychology Research Group, which has played a leading role in developing human factors research in aviation within the European Commission's 4th, 5th & 6th Framework RTD Programmes, co-ordinating and participating in a series of projects that have addressed human factors in the aircraft maintenance industry, and in the wider aviation system. Other current research addresses safety in commercial road transport and in process industries. Through the European Transport Safety Council he has also contributed to various forums and reports on transport safety policy in Europe. His main research interests are in organisational aspects of safety and efficiency and with problems of implementation, change and innovation.

Jean Pariès is CEO of the Dédale SA company (Paris, France). Dédale's activity focuses on the Human Factors dimensions of safety, across the domains of aviation, rail, nuclear power, shipping, road tunnels, and health care. Jean is also Associated Research Director with the Centre National de Recherche Scientifique (CNRS).

He graduated from ENAC, the French National School of Civil Aviation, as an aeronautical engineer in 1973. He joined the French civil aviation authority (DGAC), where he worked in a succession of roles including airworthiness regulation, operational regulation, and pilot licensing regulation. From 1988 he participated in the foundation of the ICAO Flight Safety and Human Factors Working Group chaired by Captain Daniel Maurino. In 1990, he joined the French air accident investigation body, the Bureau Enquêtes Accidents (BEA) as Deputy Head, and Head of Investigations. He led the investigation into the A320 aircraft accident at Mont Saint-Odile, France (1992). He contributed to the evolution of accident analysis methods and particularly to the introduction of an organisational perspective into the ICAO Annex 13 accident investigation guidelines. Jean left the BEA in 1994 and has, since that time, led the Dédale team on a wide variety of applied human factors, training development, operational research and organisational reliability consulting assignments. He has also taught

Human Factors, Human Reliability and Organisational Reliability for more than 15 years at several schools and universities.

Gunilla A. Sundström is Senior Vice President with Wachovia Corporation leading Wachovia's Outsourcing Governance & Program Office. In this role she is responsible for development of governance capabilities, processes and tools in support of outsourcing-related decisions at an enterprise level. For more than 18 years, she has been engaged in decision making research in complex systems in various domains including industrial process control, telecommunications and financial services. Her primary research interests include decision modeling, analysis and design of decision support for complex systems. She has authored more than 60 papers/book chapters and articles in refereed journals, technical journals and magazines. She currently serves on the editorial boards of the *International Journal of Human – Computer Studies* and the *International Journal of Cognition Technology & Work*. She has been awarded various leadership and teamwork commendations including the IEEE-Systems Man Cybernetics Society's outstanding contributions award. Dr. Sundström holds a Dr. phil. degree from University of Mannheim, Germany.

Ron Westrum is Professor of Sociology and Interdisciplinary Technology at Eastern Michigan University. A specialist in the sociology of science and technology and in systems safety, he has lectured widely to the international community in aviation, medicine, and nuclear power. He has numerous publications, including three books, the last of which is 'Sidewinder: Creative Missile Development at China Lake' (1999). He is a management consultant through his firm Aeroconcept, LLC. A member of the Human Factors and Ergonomics Society, he is writing a book on information flow. He lives in Ann Arbor, Michigan, with his wife Deborah and two children, Max and Piper.

David D. Woods is Professor at Ohio State University and past-president of the Human Factors and Ergonomics Society. From his initial work following the Three Mile Island accident in nuclear power, to studies of coordination breakdowns between people and automation in aviation accidents, to his work on to his role in today's national debates about patient safety, he has developed and advanced the foundations and practice of Cognitive Engineering. David Woods was President and is Fellow of the Human Factors and Ergonomic Society as well as Fellow of the American Psychological Society, and the American Psychological Association. He received the Kraft Innovators award from the Human Factors and Ergonomic Society for developing the foundations of Cognitive Engineering and shared the Ely Award for best paper in the journal *Human Factors* (1994), as well as a Laurels Award from *Aviation Week and Space Technology* (1995) for research on the human factors of highly automated cockpits.

John Wreathall is a Specialist in the field of human and organizational performance, who has worked for over 35 years in the fields of nuclear power, medicine, aviation and aerospace, rail transport, chemical process and pharmaceutical manufacturing. He has led the development of two state-of-the-art human reliability analysis methods, and participated in nine probabilistic safety assessments, mostly applied to nuclear power, but most recently in the fields of transplant surgery and bio-terrorism. In addition, a major activity he has led has been the development of leading indicators of organizational behaviour as a management tool for use in the nuclear and aerospace industries.

Mr. Wreathall has degrees in nuclear engineering and systems engineering from Queen Mary College, The University of London. He is president of John Wreathall & co., Inc., based in Dublin, Ohio.

Contributing Authors

Dr. Betty Barrett, MIT Research Scientist based in the Center for Technology, Policy and Industrial Development (CTPID), expertise in the application of complexity principles in organizational settings, workplace dynamics, and the operation of geographically distributed teams.

John S. Carroll is Professor of Behavioral and Policy Sciences at the MIT Sloan School of Management and the MIT Engineering Systems Division and Co-Director of the MIT Lean Aerospace Initiative. He taught previously at Carnegie-Mellon University, Loyola University of Chicago, and the University of Chicago. He received a B.S. (Physics) from MIT and a Ph.D. (Social Psychology) from Harvard. His research has focused on decision-making and learning in organizational contexts. Current projects examine organizational safety issues in high-hazard industries such as nuclear power, aerospace, and health care, including self-analysis and organizational learning, safety culture, leadership, communication and systems thinking. He has published five books and over 70 journal articles and book chapters.

Dr. Joel Cutcher-Gershenfeld is a Senior Research Scientist in MIT's Sloan School of Management. He has over 20 years experience in studying and leading large-scale systems change and institutional re-alignment initiatives. Dr. Cutcher-Gershenfeld is co-author of *Valuable Disconnects in Organizational Learning Systems* (Oxford University Press, 2005), *Lean Enterprise Value* (Palgrave/Macmillan, 2002), *Knowledge-Driven Work* (Oxford University Press, 1998), *Strategic Negotiations* (Harvard Business School Press, 1994) and three additional co-authored or co-edited books, as well as over 60 articles on large-scale systems change, new work systems, labor-management relations, negotiations, conflict resolution, organizational learning, public policy and economic development.

Nicolas Dulac is a Ph.D. student in the Aeronautics and Astronautics Department at MIT. His research interests are in safety and visualization. He is currently working on the design of advanced risk management tools.

Louis Goossens is a Chemical Engineer by training. Since 1980 he has been (senior) lecturer in Safety Science at Delft University of Technology. He specializes in methods of risk assessment and the modeling of risk, risk control and management. He has published widely on the use of expert judgment for risk assessment and managed a number of international projects, covering risks ranging from the nuclear and process industry to environmental and transport. He has also developed a number of inspection and audit systems. He is associate editor of the *Journal of Risk Research*.

Frank Guldenmund studied psychology at Leiden University and majored in cognitive psychology and methodology. Joining the Safety Science Group in 1992 primarily as a methodologist his involvement subsequently enlarged to include the modelling of safety management systems and the role of organisational culture in relation to (occupational) safety. He has been involved in projects regarding both the modelling and the measurement of these variables. Being quantitative by education his interest has gradually become more qualitative and holistic.

Tom Heijer graduated as Control Engineer from Delft University of Technology in 1975. He worked at the Dutch Institute for Road Safety Research on crash simulation and human biomechanics and later on long term safety research including the safety implications of telematics applications. Since 1989 he has worked part time at the Safety Science Group of the Delft University of Technology on a broad range of transport safety projects in the road, rail and air traffic sectors. His specialization is in system safety, risk modeling and feedback systems.

Dr. Christopher Nemeth is a Research Associate (Assistant Professor) in the Department of Anesthesia and Critical Care at the University of Chicago who studies human performance in complex high hazard environments. Recent research interests include technical work in high stakes settings, research methods in individual and distributed cognition, and understanding how information technology erodes or enhances system resilience. He has experience as a corporate manager, design and human factors consultant, university instructor, expert witness, and career active duty and reserve US Naval officer. His book on human factors research methods, *Human Factors Methods for Design*, is available from Taylor and Francis/CRC Press.

David Zipkin has a master's degree in Technology and Policy from the Massachusetts Institute of Technology. Prior to MIT, he worked in the software industry.

Bibliography

Adamski, A. & Westrum, R. (2003). Requisite imagination. The fine art of anticipating what might go wrong. In E. Hollnagel (Ed.), *Handbook of cognitive task design* (pp. 193-220). Mahwah, NJ: Lawrence Erlbaum Associates.

Ale, B. J. M., Bellamy, L. J., Cooke, R. C., Goossens, L. H. J., Hale, A. R., Roelen, A. & Smith, E. (2004). *Development of a causal model for air transport safety.* Delft University of Technology: Safety Science Group.

Amalberti, R. (2001). The paradoxes of almost totally safe transportation systems. *Safety Science, 37,* 109-126.

Amalberti, R., Auroy, Y., Berwick, D. & Barach, P. (2005). Five system barriers to achieving ultrasafe health care. *Ann. Intern. Med., 142(9),* 756-764.

Amalberti, R. & Hourlier, S. (2005). Human error reduction strategies, In P. Carayon (Ed.), *Handbook of human factors and ergonomics in healthcare and patient safety.* Hillsdale, NJ: Lawrence Erlbaum Associates.

Ashby, W. R. (1956). *An introduction to cybernetics.* London: Methuen & Co.

Auroy, Y., Amalberti, R. & Benhamou, D. (2004). Risk assessment and control require analysis of both outcomes and process of care, editorial. *Anesthesiology, 101,* 815-817.

Baker, A. (2002). New York seeking better response to catastrophes. *New York Times,* 20 August, A1 & A19.

Banbury, S. & Tremblay, S. (2004). (Eds.) *Situation awareness: A cognitive perspective.* Aldershot: Ashgate.

Barabási, A.-L. & Bonabeau, E. (2003). Scale-free networks. *Scientific American, 288,* 60-69.

Bedau, M. A. (1997). Weak emergence. *Philosophical Perspectives, 11,* 375–399.

Bedau, M. A. (2002). Downward causation and the autonomy of weak emergence. *Principia, 6,* 5–50.

Bellamy, L. J., Papazoglou, I. A., Hale, A. R., Aneziris, O. N., Ale, B. J. M., Morris, M. I. & Oh, J. I. H. (1999). *I-Risk: Development of an integrated technical and management risk control and monitoring methodology* for managing and quantifying on-site and off-site risks (Contract ENVA-CT96-0243). Report to European Union. Ministry of Social Affairs and Employment. Den Haag.

Benner, L. J. (1978). Five accident theories and their implications for research. Joint International Meeting of the American Association for Automotive Medicine and the International Association for Accident and Traffic Medicine, July 10, 1978, Ann Arbor, MI: http://www.iprr.org/THEORY/AnnArbor78.htm.

Best, M. H. (2001). The new competitive advantage. Oxford: Oxford University Press.

Biever, C. & Hecht, J. (2005). Alarm bells ringing over US response to tsunami scare. New Scientist, Issue No. 2505, 8-9.

Bigley, K. & Roberts. K. (2001). The incident command system: High-reliability organizing for complex and volatile task environments. Academy of Management Journal, 44, 1281-1299.

Boeing. How safe is flying? Website accessed 07-05-05. http://www.boeing.com/commercial/safety/pf/pf_howsafe.html

Bonabeau, E. & Théraulaz, G. (2000). Swarm smarts. Scientific American, March, 73-79.

Bonabeau, E., Dorigo, M. & Théraulaz, G. (1999). Swarm intelligence: From natural to artificial systems. Oxford University Press.

Borgenstam, C. & Sandström, A. (1995). Why Vasa capsized. Stockholm: AB Grafisk Press.

Bourrier, M. (1998). Elements for designing a self-correcting organisation: Examples from nuclear plants. In A. R. Hale & M. Baram (Eds.) Safety management: The challenge of change. Oxford: Pergamon.

Branscome, D. R. (1999). WIRE Mishap investigation board report. NASA, June 8.

Brinded, M. (2000). Perception vs. analysis. How to handle risk. The Lloyd's Register Lecture. March. London: Royal Academy of Engineering.

Cabbage, M. & Harwood, W. (2004). Comm check: The final flight of shuttle Columbia. New York: Free Press.

Carlson, J. M. & Doyle, J. C. (2000). Highly optimized tolerance: Robustness and design in complex systems. Physical Review Letters, 84(11), 2529-2532.

Carlson, J. M. & Doyle, J. C. (2002). Complexity and robustness. In: Proceedings of the A. M. Sackler Colloquium of the National Academy of Science; PNAS, 99(Suppl. 1), 2538-2545.

Carson, W. (1981). *The other price of Britain's oil. Safety and control in the North Sea.* London: Martin Robertson.

Checkland, P. (1981). *Systems thinking, systems practice.* Chichester, UK: John Wiley & Sons, Ltd.

Checkland, P. & Scholes, J. (1990). *Soft systems methodology in action.* Chichester, UK: John Wiley & Sons, Ltd.

Cook, R. I. & O'Connor, M. (2005). Thinking about accidents and systems. In H. Manasse & K. Thompson (Eds.), *Improving medication safety.* American Society for Health-System Pharmacists, Bethesda MD.

Cook, R. I. & Rasmussen, J. (2005). Going solid: A model of system dynamics and consequences for patient safety. *Quality & Safety in Health Care, 14*(2), 130-134.

Cook, R. I., Render, M. L. & Woods, D. D. (2000). Gaps in the continuity of care and progress on patient safety. *British Medical Journal, 320,* 791-794.

Cook, R. I., Woods, D. D. & McDonald, J. S. (1991). *Human performance in anesthesia: A corpus of cases.* Cognitive Systems Engineering Laboratory Report, prepared for Anesthesia Patient Safety Foundation, April 1991.

Cook, R. I., Woods, D. D. & Miller, C. (1998). *A tale of two stories: Contrasting views on patient safety.* Chicago, IL: National Patient Safety Foundation.

Corrigan, S. (2002). *Comparative analysis of safety management systems and safety culture in aircraft maintenance.* Doctoral Thesis. Trinity College Dublin: Department of Psychology.

Corrigan, S., Morrison, R. & McDonald, N. (2004). *Contributions to the safety and efficiency of the multi-drop or local/short haul operations.* Progress Report June 2004. Centre for Transportation Research and Innovation for People (TRIP). Trinity College Dublin.

Coutu, D. (2002). How resilience works. *Harvard Business Review,* May, 46-55.

Cox, S. & Flin, R. (1998). Safety culture. Philosopher's stone or man of straw? *Work & Stress, 12,* 189-201.

Crichton, M., Flin, R. & Rattray, W. (2002). Training decision makers using tactical decision games. *Journal of Contingencies and Crisis Management, 8,* 208-217.

Csete, M. E. & Doyle, J. C. (2002). Reverse engineering of biological complexity. *Science, 295,* 1664-1669.

Cullen, D. (1990). *The Piper Alpha disaster.* London: HMSO.

Cutler, T. & James, P. (1996). Does safety pay? *Work, Employment & Safety, 10,* 755-765.

Davies, P. C. W. (2004). Emergent biological principles and the computational properties of the universe. *Complexity, 10*(2), 11-15.

Deighton, L. (1996). *Fighter: The true story of the Battle of Britain.* London, UK: Pimlico.

Dekker, S. W. A. (2002). *The field guide to human error investigations.* Bedford, UK: Cranfield University Press/Aldershot, UK: Ashgate.

Dekker, S. W. A. (2003a). Illusion of an explanation. *International Journal of Aviation Psychology, 13*(2), 95-106.

Dekker, S. W. A. (2003b). When human error becomes a crime. *Journal of Human Factors and Aerospace Safety, 3*(1), 83-92.

Dekker, S. W. A. (2005). *Ten questions about human error: A new view of human factors and system safety.* Hillsdale NJ: Lawrence Erlbaum.

Diamond, J. (2005). *Collapse. How societies choose to fail or survive.* London, UK: Allen Lane.

Dijkstra, G. & van Poortvliet, A. (2004). *Veiligheid: Een veilige balans met kosten* (Safety: A safe balance with costs). Essay. Prorail Strategy & Innovation. Utrecht.

Dominguez, C. O., Flach, J. M., McDermott, P. L., McKellar, D. M. & Dunn, M. (2004). The conversion decision in laparoscopic surgery: Knowing your limits and limiting your risks. In K. Smith, J. Shanteau & P. Johnson (Eds.), *Psychological investigations of competence in decision making* (pp. 7-39). Cambridge: Cambridge University Press.

Drogoul, F., Grau, J. Y. & Amalberti, R. (2003). *Dialogues with the customer: The designer interface with operations personnel.* Paper to the New Technology & Work Symposium, Blankensee. Bretigny sur Orge: Eurocontrol.

Duijm, N.-J., Hale, A. R., Goossens, L. H. J. & Hourtolou, D. (2004). Evaluating and managing safety barriers in major hazard plants. In C. Spitzer, U. Schmocker & V. N. Dang (Eds.), *Probabilistic safety assessment & management.* Berlin: Springer. 110-115.

Dupuy, J. P. & Teubner, G. (1990). *Paradoxes of self references in the humanities, law, and the social sciences.* Stanford: Anma Libri.

Dupuy, J. P. (2002). *Pour un catastrophisme éclairé.* Paris: Seuil.

Dwyer, J. (2002). A calamity unimaginable in scope, and unexamined in all its dimensions. *New York Times,* 12 September.

Elling, M. G. M. (1991). *Veiligheidsvoorschriften in de industrie* (Safety rules in industry). PhD Thesis. University of Twente, Netherlands: Faculty of Philosophy and Social Sciences Publication WMW No.8.

Endsley, M. & Garland, D. (2002). (Eds.) *Situation awareness.* Mahwah, NJ: Lawrence Erlbaum Associates.

European Transport Safety Council. (2003). *Transport safety performance in the EU: A statistical overview.* Brussels.

Fields, C. (2002). Army delayed Joola Rescue Bid. *Lloyds List, 8* November.

Fischhoff, B. (1975). Hindsight ≠ foresight: The effect of outcome knowledge on judgment under uncertainty. *Journal of Experimental Psychology: Human Perception and Performance, 1,* 288-299.

Fiterman, C. (2003). Prévention et gestion des risques technologiques et industriels. *Avis et rapport du Conseil Economique et Social, Editions des Journaux officiels, 7,* 41103-0007.

Flin, R. (1996). *Sitting in the hot seat. Leaders and teams for critical incident management.* Chichester: Wiley.

Flin, R. (2003). Men behaving badly? Senior managers and safety. *Human Factors and Ergonomics in Manufacturing, 13*(4), 1-8.

Flin, R., Mearns, K., Fleming, M. & Gordon, R. (1996). *Risk perception and safety in the offshore oil and gas industry.* (HSE Research Report OTH 94454). London: HSE Books.

Flin, R., O'Connor, P. & Crichton, M. (in prep). *Safety at the sharp end: A guide to non-technical skills.* Aldershot: Ashgate.

Flin, R. & Slaven, G. (1994). *The selection and training of offshore installation managers for crisis management.* London: HSE Books.

Free, R. (1994). *The role of procedural violations in railway accidents.* PhD thesis. University of Manchester.

Furuta, K., Sasou, K., Kubota, R., Ujita, H., Shuto, Y. & Yagi, E. (2000). *Human factor analysis of JCO criticality accident.* Division of Human-Machine Systems Studies, Atomic Energy Society of Japan.

Galea, E. R. (2001). Simulating evacuation and circulation in planes, trains, buildings and ships using the EXODUS software. In M. Schreckenberg & S. D. Sharma (Eds.), *Proceedings of the International Conference on Pedestrian and Evacuation Dynamics.* Berlin: Springer.

Gehman, H. (2003). *Columbia accident investigation report.* Washington DC: U.S. Government Accounting Office, August. Available from www.caib.us/news/report/default.html.

Girard, R. (1961). *Mensonge romantique et vérité Romanesque.* Paris: Grasset.

Global Aviation Information Network (2003). *Role of Analytical Tools in Airline Flight Safety Management* GAIN WG B in June 2003. www.gainweb.org, accessed on 01-04-05.

Gould, S. J. (1981). *The mismeasure of man.* New York: Norton.

Greenwood, M. & Woods, H. M. (1919). *The incidence of industrial accidents upon individuals with special reference to multiple accidents* (British Industrial Fatigue Research Board, Report No. 4). London: HMS.

Griffiths, P. & Stewart, S. (2005). *Presentations on Easyjet risk management programme.* Trinity College Dublin.: Department of Psychology.

Guldenmund, F. W. & Hale, A. R. (2004). *The ARAMIS audit manual. Version 1.2.* Delft University of Technology: Safety Science Group.

Haddon, Jr. W. (1966). The prevention of accidents. In: *Preventive medicine,* p. 591-621. Boston: Little Brown.

Hale, A. R. (1989). Safety of railway personnel: An outsider's view. Paper to the *International Jubilee Congress of the Dutch Railways,* 28-30 June, Amsterdam. Delft University of Technology: Safety Science Group.

Hale, A. R. (2000). The Tokai-mura accident. *Cognition Technology & Work, 2*(4), 215-217.

Hale, A. R. (2001). Regulating airport safety: the case of Schiphol. *Safety Science, 37*(2-3), 127-149.

Hale, A. R., Goossens, L. H. J., Ale, B. J. M., Bellamy, L. A., Post, J., Oh, J. I. H. & Papazoglou, I. A. (2004). Managing safety barriers and controls at the workplace. In C. Spitzer, U. Schmocker & V. N. Dang (Eds.), *Probabilistic Safety Assessment & Management,* pp. 608-613. Berlin: Springer Verlag.

Hale, A. R., Guldenmund, F. W., Goossens, L. H. J., Karczewski, J., Duijm, N.-J., Hourtolou, D., LeCoze, J.-C., Plot, E., Prats, F., Kontic, B., Kontic, D. & Gerbec, M. (2005). Management influences on major hazard prevention: The ARAMIS audit. Paper to the *ESREL conference on Safety & reliability,* Gdynia.

Hale, A. R., Heijer, T. H. & Koornneef, F. (2002). *Management van Veiligheidsregels bij de Nederlandse Spoorwegen.* (Management of safety rules by the Dutch Railways) Report to Railned. Delft University of Technology: Safety Science Group.

Hale, A. R., Heijer, T. H. & Koornneef, F. (2003). Management of safety rules: The case of railways. *Safety Science Monitor, 7*(1). http://www.ipso.asn.au/vol7/vol7idx.htm.

Hallahan, W. H. (1994). *Misfire: The history of how America's small arms have failed our military*. New York: Scribners.

Hammond, P. & Mosley, M. (2002). *Trust Me I'm a Doctor*. London: Metro.

Hauser, C. (2004). At US hospital, Reflections on 11 hours and 91 casualties. *New York Times*, Dec. 29.

Heijer, T. H. & Hale, A. R. (2002). *Risk knows no bounds: How rational can we be?* Paper to the New Technology & Work Workshop, Blankensee. Delft University of Technology: Safety Science Group.

Heinrich, H. W. (1931). *Industrial accident prevention*. New York: McGraw-Hill.

Heinrich, H. W., Petersen, D. & Roos, N. (1980). *Industrial accident prevention* (Fifth Edition). New York: McGraw-Hill Book Company.

Helmreich, R. L. (2001). On error management: Lessons from aviation. *British Medical Journal, 320*, 781-784.

Helmreich, R. L. & Sexton, J. (2004). Managing threat and error to increase safety in medicine. In R. Dietrich & K. Jochum (Eds.), *Teaming up: Components of safety under high risk*. Aldershot: Ashgate.

Hoekstra, J. M. (2001). *Designing for safety: The free flight air traffic management concept*. PhD. Thesis. Delft University of Technology: Safety Science Group.

Hofmann, D. & Morgeson, F. (2004). The role of leadership in safety. In J. Barling & M. Frone (Eds.) *The psychology of workplace safety*. Washington: APA Books.

Holland, J. H. (1998). *Emergence*. Reading, MA: Helix.

Hollnagel, E. (1983). *Position paper on human error*. NATO Conference on Human Error, August 1983, Bellagio, Italy. http://www.ida.liu.se/~eriho/, accessed on 2005-06-15.

Hollnagel, E. (1993a). The phenotype of erroneous actions. *International Journal of Man-Machine Studies, 39*, 1-32.

Hollnagel, E. (1993b). *Human reliability analysis: Context and control*. Academic Press, London.

Hollnagel, E. (1998). Cognitive reliability and error analysis method (CREAM). New York: Elsevier Science Inc.

Hollnagel, E. (2004). *Barriers and accident prevention*. Aldershot, UK: Ashgate.

Hollnagel, E. (2005). *The natural unnaturalness of rationality*. Proceedings of 7th International Conference on Naturalistic Decision Making, 15-17 June 2005, Amsterdam, The Netherlands.

Hollnagel, E. & Woods, D. D. (2005). *Joint cognitive systems: Foundations of cognitive systems engineering.* Boca Raton, FL: Taylor & Francis/CRC Press.

Holt, L. T. C. (1955). *Red for danger* (4th Edition). Newton Abbott. David & Charles.

Hopkins, A. (2000). *Lessons from Longford.* Sydney: CCH.

Hopkins, A. (2005). *Safety, Culture and Risk. The Organisational Causes of Disasters.* Sydney: CCH.

Houtenbos, M., Hagenzieker, M., Wieringa, P. & Hale, A. R. (2004). *The role of expectations in interaction behaviour between car drivers.* Delft University of Technology: Safety Science Group.

Hudson, P. (1999). *Safety culture – the way ahead? Theory and practical principles.* Leiden University. Leiden:Centre for Safety Science.

Hughes, T. (1983). *Networks of power electrification in western society.* Baltimore: John Hopkins University Press.

Hughes, T. (1994). *System momentum and system adaptation* (Personal communication).

Hutchins, E. (1996). *Cognition in the wild.* Cambridge, MA: MIT Press.

Hutter, B. (2001). *Regulation and risk.* Oxford: Oxford University Press.

ICAO (2001). *Annex 13: Aircraft accident and incident investigation* (Ninth edition).

Institute of Medicine (1995). *HIV and the blood supply: An analysis of crisis decision-making.* Washington DC: National Academies Press.

Itoh, K., Andersen, H. B. & Seki, M. (2004). Track maintenance train operators attitudes to job, organisation and management, and their correlation with accident/incident rate. *Cognition, Technology & Work, 6*(2), 63-78.

Jackson, R. & Watkin, C. (2004). The resilience inventory: Seven essential skills for overcoming life's obstacles and determining happiness. *Selection and Development Review, 20,* 6, 11-15.

Janis, I. (1982). *Groupthink: Psychological study of policy decisions and fiascos.* Boston: Houghton Mifflin.

Johnson, P. E., Jamal, K. & Berryman, R. G. (1991). Effects of framing on auditor decisions. *Organizational Behavior and Human Decision Processes, 50,* 75-105.

Johnson, W. G. (1980). *MORT safety assurance systems.* New York: Marcel Dekker, Inc.

Kennedy, I. (2001). Learning from Bristol. The report of the public inquiry into children's heart surgery at the Bristol Royal Infirmary 1984-1995. *Command Paper CM 5207*. London: HMSO.

Kennis, P. & Marin, B. (Eds.) (1997). *Managing AIDS: Organizational responses in six European countries*. Aldershot: Ashgate.

Kessler, E., Bierley, P. & Gopalakrishnan, S. (2001). Vasa syndrome: Insights from a 17th century new product disaster. *Academy of Management Executive, 15*, 80-91.

Klampfer, B., Flin, R., Helmreich, R. L., Hausler, R., Sexton, B., Fletcher, G., Field, P., Staender, S., Lauche, K., Dieckmann, P. & Amacher, A. (2001). Enhancing performance in high risk environments: Recommendations for the use of behavioural markers. *Behaviour Markers Workshop*, Zurich, July 5-6.

Klein, G., Pliske, R., Crandall, B. & Woods, D. D. (2005). Problem detection. *Cognition, Technology & Work, 7*(1), 14-28.

Klinect, J. R., Murray, P., Merritt, A. C. & Helmreich, R. L. (2003). Line operations safety audit (LOSA). Definition and operating characteristics. In proceedings of the *12th International Symposium on Aviation Psychology* (pp. 663-668). Dayton, OH: The Ohio State University.

Kohn, L., Corrigan, J. & Donaldson, M. (1999). *To err is human - building a safer health system*. Committee on Quality in America. Washington DC, Institute of Medicine. National Academic Press.

Kolata, G. (2004). Program coaxes hospitals to see treatments under their noses. *New York Times, Dec. 25*, A1, C6.

Kranz, G. (2000). *Failure is not an option: Mission control from Mercury to Apollo 13 and beyond*. New York: Simon and Schuster.

Kube, C. R. & Bonabeau, E. (2000). Cooperative transport by ants and robots. *Robotics and Autonomous Systems, 30*(1-2), 85-101.

Kuhn, T. S. (1970). *The structure of scientific revolutions* (2nd Ed.). Chicago: University of Chicago Press.

Kurpianov, A. (1995). Derivatives debacles. *Economic Quarterly* (Federal Reserve Bank of Richmond), *81*, 1-24.

Lanir, Z. (1986). *Fundamental surprise: The national intelligence crisis*. Eugene, OR: Decision Research. (originally Tel Aviv: HaKibbutz HaMeuchad, 1983, Hebrew).

Laplace, P. S. (1814). *Essai philosophique sur le fondement des probabilités*. C. Bourgeois, Paris, 1986; Philosophical Essays on Probabilities. Translated by Truscott, F. W.; Emory, F. L. Dover, New York.

Larsen, L., Petersen, K., Hale, A. R., Heijer, T. H., Harms Ringdahl, L., Parker, D. & Lawrie, D. (2004). *A Framework for safety rule management* (D2.8.1). SAMRAIL Final report Contract No. GMA2/2001/52053. Lyngby. Danish Traffic Institute.

Lawrence, P. R. & Dyer, D. (1983). *Renewing American industry.* The Free Press: New York.

LeMessurier, W. (2005). *The fifty-nine story crisis. A lesson in professional behavior.* http://onlineethics.org/moral/lemessurier. Accessed on 1/11/05.

Leplat, J. (1987). Occupational accident research and systems approach in Rasmussen, J., Duncan, K. & Jacques Leplat, J. (Eds.), *New Technology and Human Error,* pp. 181-191, New York: John Wiley & Sons.

Leveson, N. G. (2002). *A new approach to system safety engineering.* Cambridge, MA: Aeronautics and Astronautics, MIT.

Leveson, N. G. (2004). A New Accident Model for Engineering Safer Systems. *Safety Science, 42*(4), 237-270.

Leveson, N. G., Cutcher-Gershenfeld, J., Dulac, N. & Zipkin, D. (2005). *Phase 1 Final Report on Modeling, Analyzing and Engineering NASA's Safety Culture.* http://sunnyday.mit.edu/Phase1-Final-Report.pdf. [NL].

Leveson, N. G., Daouk, M., Dulac, N. & Marais, K. (2003). Applying STAMP in accident analysis. *Workshop on the Investigation and Reporting of Accidents,* Sept.

Link, D. C. R. (2000). *Report of the Huygens Communications System Inquiry Board,* NASA, December 2000.

Luthans, F. & Avolio, B. (2003). Authentic leadership development. In K. Cameron, J. Dutton & R. Quinn (Eds.) *Positive organizational scholarship.* San Francisco: Berrett-Koehler.

Marais, K. & Leveson, N. G. (2003). Archetypes for Organizational Safety, *Workshop on the Investigation and Reporting of Accidents,* Sept.

Mattila, M., Hyttinen, M. & Ratanen, E. (1994). Effective supervisory behaviour and safety at the building site. *International Journal of Industrial Ergonomics, 13,* 85-93.

McCurdy, H. (1993). *Inside NASA: High technology and organizational change in the U.S. space program,* Johns Hopkins University Press.

McDonald, H. (2000). *Shuttle independent assessment team (SIAT) report,* NASA, February.

McDonald, N. (Ed.) (1999). *Human-centred management for aircraft maintenance.* Deliverable report to the European Commission no. ADAMS-WP4A-D2 ADAMS Project N° BE95-1732. Trinity College Dublin: Department of Psychology.

McDonald, N., Corrigan, S. & Ward, M. (2002). *Well-intentioned people in dysfunctional systems.* Keynote paper presented at the 5th Workshop on Human error, safety and systems development, Newcastle, Australia.

McDonald, N., Ward, M., Corrigan, S., Mesland, M. & Bos, T. (2001). *AMPOS evaluation – Pilot implementation.* Deliverable to the European Commission no. 7.2. AMPOS Project N° 29053 (Esprit). Trinity College Dublin: Department of Psychology.

McLennan, J., Holgate, A. M., Omodei, M. M. & Wearing, A. J. (2005). *Decision making effectiveness in wildfire incident management teams.* Proceedings of 7th International Conference on Naturalistic Decision Making, 15-17 June 2005, Amsterdam, The Netherlands.

Mearns, K., Flin, R.., Fleming, M. & Gordon, R. (1997). *Human and organisational factors in offshore safety.* (HSE Research Report OTH 543). London: HSE Books.

Mearns, K., Flin, R., Gordon, R. & Fleming, M. (1998). Measuring safety climate in the offshore oil industry. *Work & Stress, 12,* 238-254.

Mearns, K., Whitaker, S. & Flin, R. (2003). Safety climate, safety management practice and safety performance in offshore environments. *Safety Science, 41,* 641-680.

Merton, R. K. (1936). The unanticipated consequences of social action. *American Sociological Review, 1,* 894-904.

Mitroff, I. (1974). On systemic problem solving and the error of the third kind. *Behavioral Science, 19*(6), 383-393.

Morgan, G. (1986). *Images of organization.* Newbury Park, CA: Sage.

Morowitz, H. J. (2002). *The emergence of everything;* Oxford University Press, Oxford.

NASA/ESA Investigation Board (1998). *SOHO Mission interruption,* NASA, 31 August.

National Commission on Terrorist Attacks. (2004). *9/11 Report.* Scranton, PA: W. W. Norton.

Nemeth, C. (2003). *The master schedule: How cognitive artifacts affect distributed cognition in acute care.* Dissertation Abstracts International 64/08, 3990. (UMI No. AAT 3101124).

Nemeth, C., Cook, R. I. & Woods, D. D. (2004). Messy details: Insights from the study of technical work in healthcare. *IEEE SMC Part A*, *34*(6), 689-692.

Nichols, T. (1997). *The sociology of industrial injury*. London: Mansell.

Nonaka, I. (2002). A dynamic theory of organizational knowledge creation. In C. Choo & N. Bontis (Eds.), *The strategic management of intellectual capital and organizational knowledge*. Oxford: Oxford University Press (pp. 437-462).

Norman, D. A. (1990). The 'problem' with automation: Inappropriate feedback and interaction, not 'over-automation'. *Philosophical Transactions of the Royal Society of London, B 327*, 585-593.

NRC. (2000). *Technical basis and implementation guidelines for a technique for human event analysis* (ATHEANA) (NUREG-1624, Rev. 1). Rockville, MD: U.S. Nuclear Regulatory Commission.

NTSB. (1983). *Accident investigation symposium* (Report RP-84/01). Springfield, Virginia April 26-28, 1983.

NTSB. (2002). http://www.ntsb.gov/Pressrel/2002/table01.pdf website accessed on 01-04-05.

Ohrelius, B. (1962). *Vasa, The King's ship*. (Transl. Maurice Michael). London: Cassell.

Ostrom, E. (1990). *Governing the commons: The evolution of institutions for collective action*. New York: Cambridge University Press.

Ostrom, E. (1999). Coping with tragedies of the commons. *Annual Review of Political Science, 2*, 493-535.

Ostrom, E. (2003). Toward a behavioral theory linking trust, reciprocity, and reputation. In E. Ostrom & J. Walker (Eds.), *Trust and reciprocity: Interdisciplinary lessons from experimental research*. New York: Russell Sage Foundation.

Pate-Cornell, E. (1990). Organizational aspects of engineering system safety: The case of offshore platforms. *Science, 250*, 1210-1217.

Patterson, E. S. & Woods, D. D. (2001). Shift changes, updates, and the on-call model in space shuttle mission control. *Computer supported cooperative work. The Journal of Collaborative Computing, 10*(3-4), 317-346.

Patterson, E. S., Cook, R. I., Woods, D. D. & Render, M. L. (2004a). Examining the complexity behind a medication error: Generic patterns in communication. *IEEE SMC Part A, 34*(6), 749-756.

Patterson, E. S., Roth, E. M., Woods, D. D., Chow, R. & Gomes, J. O. (2004b). Handoff strategies in settings with high consequences for failure: Lessons for health care operations. *International Journal for Quality in Health Care, 16*(2), 125-132.

Pérezgonzález, J. D. (2004). *Construction safety: A systems approach.* Doctoral dissertation. University of La Laguna, Tenerife, Spain.

Pérezgonzález, J. D., Ward, M., Baranzini, D., Liston, P. & McDonald, N. (2004). *Organisational and operational performance assessment.* Deliverable to the European Commission no. 3.2.1, ADAMS-2 project no. GRD1-2000-25751. Trinity College Dublin: Department of Psychology.

Perrow, C. (1984). *Normal accidents: Living with high risk technologies.* New York: Basic Books, Inc.

Pettigrew, A. & Whipp, R. (1991). *Managing change for competitive success.* Cambridge (Mass). Blackwell.

Pilon, N., Kirwan, B., Hale, A. R. & Pariés, J. (2001). *EEC safety research & development strategy: Output of the "Comite des Sages"* (Committee of wise men). Bretigny s/Orge: Eurocontrol.

Pliske, R., Crandall, B. & Klein, G. (2004). Competence in weather forecasting. In K. Smith, J. Shanteau & P. Johnson (Eds.), *Psychological investigations of competence in decision making* (pp. 7-39). Cambridge: Cambridge University Press.

Polet, P., Vanderhaegen, F. & Amalberti, R. (2003). Modelling the border-line tolerated conditions of use. *Safety Science, 41*(1), 111-136.

Porter, M. E. (1998). Clusters and the new economics of competition. *Harvard Business Review*, reprint 98609.

Prorail (2003). *Prorail aan het werk in 2003: Jaarbericht 2003* (Prorail at work in 2003: Annual report 2003). Utrecht Prorail – www.prorail.nl.

Quinn, R. (1988). *Beyond Rational Management: Mastering the Paradoxes and Competing Demands of High Performance.* San Francisco: Jossey Bass.

Raad voor de Transportveiligheid (Dutch Transport Safety Board) (2003). *Botsing in the lucht tussen de vliegtuigen PH-BWC en PH-BWD van de KLM Luchtvaartschool, beiden van het type Beech Bonanza A36, nabij Smilde op 8 Juni 2000 (Mid-air collision between aircraft PH-BWC and PH-BWD of the KLM Flight Academy, both Beech Bonanza A36's, near Smilde (Netherlands), 8 June 2000.* The Hague, NL: Author.

Rasmussen J. (1983). *Position paper for NATO Conference on Human Error.* August 1983, Bellagio, Italy.

Rasmussen, J. (1986). *Information processing and human-machine interaction: An approach to cognitive engineering.* New York: Elsevier Science Publishing.

Rasmussen J. (1990). The role of error in organizing behavior. *Ergonomics, 33,* 1185-1199.

Rasmussen, J. (1997). Risk Management in a Dynamic Society, *Safety Science,* 27(2), 183-213.

Rasmussen, J., Pejtersen, A. M. & Goodstein, L. P. (1994). *Cognitive systems engineering.* New York: John Wiley and Sons.

Rasmussen, J. & Svedung, I. (2000). *Proactive risk management in a dynamic society.* Karlstad, Sweden: Swedish Rescue Services Agency.

Reason, J. T. (1990). *Human error.* Cambridge: Cambridge University Press.

Reason, J. T. (1997). *Managing the risks of organizational accidents.* Aldershot, UK: Ashgate Publishing Limited.

Reason, J. T., Parker, D. & Lawton, R. (1998). Organisational controls and safety: The varieties of rule-related behaviour. *J. Occupational & Organisational Psychology, 71,* 289-304.

Reivich, K. & Shatte, A. (2002). *The resilience factor. Seven essential skills for overcoming life's inevitable obstacles.* New York: Bantam Books.

Ringstad, A. J. & Szameitat, S. A. (2000). Comparative Study of Accident and Near Miss Reporting Systems in the German Nuclear Industry and the Norwegian Offshore Industry. *Proceedings of the Human Factors and Ergonomics Society 44th annual meeting,* 380-384.

Rochlin, G. I. (1999). Safe operation as a social construct. *Ergonomics,* 42(11), 1549-1560.

Rogers, W. R. (1987). *The Rogers' Commission Report on the Space Shuttle Challenger Accident,* U.S. Government Accounting Office.

Sarter, N. B., Woods, D. D. & Billings, C. (1997). Automation Surprises. In G. Salvendy (Ed.), *Handbook of human factors/ergonomics* (Second Ed). New York: Wiley, pp. 1926-1943.

Schupp, B., Hale, A. R., Pasman, H. J., Lemkovitz, S. M. & Goossens, L. H. J. (2005). Design for safety: Systematic support for the integration of risk reduction into chemical process design. *Safety Science* – in press.

Seligman, M. (2003). *Authentic happiness: Using the new positive psychology to realise your potential for lasting fulfilment.* New York: Nicholas Brearley.

Senge, P. M. (1990). *The fifth discipline: The art and practice of the learning organization.* New York: Doubleday Currency.

Sheridan, T. B. (1992). *Telerobotics, automation and human supervisory control.* MIT Press, Cambridge, MA.

Shrivastava, P. (1987). *Bhopal: Anatomy of a crisis.* Cambridge: Ballinger.

Smith, P. J., McCoy, E. & Orasanu, J. (2004). Distributed cooperative problem-solving in the air traffic management system. In Klein, G. & Salas, E. (Eds.), *Naturalistic decision making.* Mahwah, NJ: Erlbaum, 369-384.

Spear, A. (2000). *NASA FBC Task Final Report,* NASA, March, 2000. url: http://appl.nasa.gov/pdf/47305main_FBCspear.pdf accessed 4-27-05.

Stephenson, A. G. (1999). *Mars Climate Orbiter: Mishap investigation board report,* NASA, November 10.

Stephenson, A. G., Lapiana, L., Mulville, D., Rutledge, P., Bauer, F., Folta, D., Dukeman, G., Sackheim, R. and Norvig, P. (2000). *Report on Project Management in NASA by the Mars Climate Orbiter Mishap Investigation Board.* NASA, March 13.

Sterman, J. D. (2000). *Business dynamics. Systems thinking and modeling for a complex world.* McGraw Hill: New York.

Sundström, G. A. & Deacon, C. C. (2002). The evolution of possibility: From organizational silos to global knowledge networks. *Proceedings of the 11ᵗʰ European Conference on Cognitive Ergonomics,* Catania, Italy, September 8-11, 129-135.

Sundström, G. A. & Hollnagel, E. (2004). Operational risk Management in financial services: A complex systems perspective. Proceedings of 9ᵗʰ IFAC/IFORS/IEA Symposium *Analysis, Design, and Evaluation of Human-Machine Systems.* Atlanta, GA, USA, September 7-9.

Suparamaniam, N. & Dekker, S. W. A. (2003). Paradoxes of power: The separation of knowledge and authority in international disaster relief work. *Disaster Prevention and Management, 12*(4), 312-318.

Sutcliffe, K. M. & Vogus T. J. (2003). Organizing for resilience. In K. S. Cameron, I. E. Dutton & R. E. Quinn (Eds.), *Positive organizational scholarship.* San Francisco: Berrett-Koehler, pp. 94-110.

Thompson, P. & McHugh, D. (1995). *Work organisations.* Basingstoke: Macmillan.

Tissington, P. & Flin, R. (in press). Dynamic decision making on the fireground. *Risk Management.*

VACS. (2003). *Causale modelering en groepsrisico* (Causal modelling and group risk) Advice to the Minister. Veiligheids Advies Commissie Schiphol. Hoofddorp. www.vacs.nl.

Vaughan, D. (1996). *The Challenger launch decision: Risky technology, culture, and deviance at NASA.* Chicago: University of Chicago Press.

Vernez, D., Heijer, T. H., Wiersma, J. W. F., Rosmuller, N. & Hale, A. R. (1999). *Generic safety balance of the tunnel in the Betuweroute freightline.* Report to Managementgroep Betuwelijn. Delft University of Technology: Safety Science Group.

Visser, J. P. (1998). Developments in HSE management in oil and gas exploration and production. In A. R. Hale & M. Baram (Eds.) *Safety management: The challenge of organisational change.* Oxford, UK: Pergamon.

von Bertalanffy, L. (1952). *Problems of life. An evaluation of modern biological and scientific thought.* Harper & Brothers: New York.

von Bertalanffy, L. (1975). *Perspectives on general system theory.* George Braziller: New York.

Ward, M., Baranzini, D., Liston, P. & McDonald, N. (2004). *Case studies on the implementation of the tools.* Deliverable to the European Commission no. 3.2.2, ADAMS-2 project no. GRD1-2000-25751. Trinity College Dublin: Department of Psychology.

Weick, K. E. (1995). *Sensemaking in organizations.* Thousand Oaks: Sage.

Weick, K. E. (2001). *Making sense of the organization.* Oxford: Blackwell Business.

Weick, K. E. & Roberts, K. H. (1993). Collective mind in organizations: Heedful interrelating on flight decks. *Administrative Science Quarterly 38,* 357-381.

Weick, K. E. & Sutcliffe, K. M. (2001). *Managing the Unexpected.* San Francisco: Jossey Bass.

Weick, K. E., Sutcliffe, K. M. & Obstfeld, D. (1999). Organizing for high reliability: Processes of collective mindfulness. *Research in Organizational Behavior, 21,* 13-81.

Welchman, G. (1982). *The hut six story: Breaking the Enigma codes.* Harmondsworth: Penguin.

Westrum, R. (1991). Cultures with requisite imagination. In: Wise, J., Stager, P. & Hopkin, J. (Eds.) *Verification and validation in complex man-machine systems.* Berlin: Springer-Verlag.

Westrum, R. (2003). Removing Latent Pathogens, *6th Australian Aviation Psychologists' Conference,* Sydney.

Westrum, R. & Adamski, A. (1999) Organizational factors associated with safety and mission success in aviation environments. In Garland, J., Wise J. A. and Hopkin, V. D. (Eds.), *Handbook of aviation human factors.* Mahwah, NJ: Lawrence Erlbaum.

Wiener, E., Kanki, B. & Helmreich, R L. (1993). (Eds.) *Cockpit resource management.* San Diego: Academic Press.

Williams, J. C. (2000). New frontiers for regulatory interaction within the UK nuclear industry. Paper to the 19th NeTWork Workshop *Safety regulation: The challenge of new technology & new frontiers.* Bad Homburg.

Woods, D. D. (2000). Designing for resilience in the face of change and surprise: Creating safety under pressure. Plenary Talk, *Design for Safety Workshop*, NASA Ames Research Center, October 10, 2000.

Woods, D. D. (2003). *Creating foresight: How resilience engineering can transform NASA's approach to risky decision making.* Testimony on The Future of NASA to Senate Committee on Commerce, Science and Transportation, Washington D.C., October 29, http://csel.eng.ohio-state.edu/woods/news/woods_testimony.pdf

Woods, D. D. (2005a). Creating foresight: Lessons for resilience from Columbia. In M. Farjoun and W. H. Starbuck (Eds.), *Organization at the limit: NASA and the Columbia disaster.* Blackwell.

Woods, D. D. (2005b). Conflicts between learning and accountability in patient safety. *DePaul Law Review. 54*(2), 485-502.

Woods, D. D. & Cook, R. I. (2002). Nine steps to move forward from error. *Cognition, Technology & Work, 4*(2), 137-144.

Woods, D. D., Johannesen, L. J., Cook, R. I. & Sarter, N. B. (1994). *Behind human error: Cognitive systems, computers, and hindsight.* Wright-Patterson Air Force Base, OH: Crew System Ergonomics Information Analysis Center.

Woods, D. D., Roth, E. M. & Bennett, K. B. (1990). Explorations in joint human-machine cognitive systems. In S. Robertson, W. Zachary & J. Black (Eds.), *Cognition, computing and cooperation.* Norwood, NJ: Ablex Publishing.

Woods, D. D. & Shattuck, L. G. (2000). Distant supervision-local action given the potential for surprise. *Cognition, Technology & Work, 2*, 86-96.

Wreathall, J. (1998). Development of leading indicators of human performance. Paper presented at the *Human Performance/Root Cause/Trending (HPRCT) Workshop*, San Antonio, TX.

Wreathall, J. (2001). Systemic safety assessment of production installations. Paper presented at the *World Congress: Safety of Modern Technical Systems*, Saarbrucken, Germany.

Wreathall, J. & Merritt, A. C. (2003). Managing human performance in the modern world: Developments in the US nuclear industry. In G. Edkins & P. Pfister (Eds.), *Innovation and consolidation in aviation*. Aldershot (UK) and Burlington, VT: Ashgate.

Wright, C. (1986). Routine deaths. Fatal accidents in the offshore oil industry. *Sociological Review, 34*, 265-289.

Yamamori, H. & Mito, T. (1993). Keeping CRM is keeping the flight safe. In: E. L. Wiener, B. G. Kanki & R. L. Helmreich (Eds.), *Cockpit resource management* (p. 400). San Diego, CA: Academic Press, Inc.

Yonay, E. (1993). *No margin for error, The making of the Israeli air force*. New York: Pantheon, p. 345.

Young, K. (1966). *Churchill and Beaverbrook*. New York: Heineman.

Young, T. (2000). *Mars Program Independent Assessment Team Report*, NASA, March 14.

Yukl, G. & Falbe, C. (1992). Influence tactics and objectives in upward, downward, and lateral influence attempts. *Journal of Applied Psychology, 75*, 132-140.

Yule, S., Flin, R. & Bryden, R. (under review). *Senior managers and safety leadership*.

Zohar, D. & Luria, G. (2004). Climate as a social-cognitive construction of supervisory safety practices: Scripts as proxy of behavior patterns. *Journal of Applied Psychology, 89*, 322-333.

Zohar, D. (2003). The influence of leadership and climate on occupational health and safety. In D. Hofmann & L. Tetrick (Eds.) *Health and safety in organizations. A multilevel perspective*. San Francisco: Jossey Bass.

Author Index

Subject Index

Printed in the United States
by Baker & Taylor Publisher Services